Praise for *C*

'This book will change the way you see the w
reads like a thriller, full of stories and facts that would be beyond ~~....~~
exceptional and rigorous journalism throughout. This is a very rare and important book,
and it's also really, really funny. You will not be able to put it down'
Chris van Tulleken, bestselling author of *Ultra-Processed People*

'As alarming as it is entertaining, this brilliant book exposes a pernicious industry that has
us all in its grip. Saabira Chaudhuri does for plastics what Chris van Tulleken has done for
ultra-processed foods – first make us angry, then make us want to do something'
Hugh Fearnley-Whittingstall, host of *War on Plastic with Anita and Hugh*

'In *Consumed*, Saabira Chaudhuri lifts the lid on petrochemical and consumer goods
companies' plastic obsession – exposing how their profits come at the planet's expense.
With sharp research and a gripping story, she lays out the real cost of our throwaway
culture and challenges us to kick the plastic habit'
Ben Cohen and Jerry Greenfield, co-founders of Ben & Jerry's

'This eye-popping, engaging and rigorous book lifts the lid on the huge and rapidly rising
cost of plastic in our lives. It exposes the pivotal role that deception has played over the
decades through well-funded lobbying of government and deliberate misinforming of
consumers. The human stories of people trying to do the right thing, or of convincing
themselves their industry is not as bad as it seems, are as compassionate as they are critical.
There are others, of course, who put short-term profit blatantly above human and planetary
health – *Consumed* gives us the good, the bad and the ugly of our plastic history. With
global plastic consumption and waste forecast to triple by 2060, this timely book calls for
a much-needed reset on plastic and lays bare the myths of a recycling future. This book is
for anyone who cares about the health of people or planet, and wants to know what to do'
Mike Berners-Lee, author of *A Climate of Truth* and *There is No Planet B*

'If you're a CEO in an industrial complex like Big Plastic, you'd better hope that Saabira
Chaudhuri doesn't turn her interest your way. As she does in *Consumed*, she'll pry loose
your playbook and lay bare the strategy that's common to so many companies that sell us
goods with huge hidden costs: the denial, the delay, the pretence of giving in, with some
secretly funded science to confuse us all (her chapter on throw-away diapers is a gem). A
must read for anyone who buys anything plastic'
Michael Moss, Pulitzer Prize-winning author of *Salt, Sugar, Fat*

'*Consumed* examines the rise and rise of plastic consumption – our addiction to the
material as users, and the role big business plays as pusher. Chaudhuri's book is convincing
and expertly written, driven by impressively deep research and a sharp understanding of
how plastic consumption illuminates many of the bigger issues that underpin consumer
behaviour, from widespread greed to a convenience-above-all mentality that often trumps
safety and environmental awareness. An important and engaging read'
Adam Alter, bestselling author of *Irresistible* and *Anatomy of a Breakthrough*

'*Consumed* tells the terrifying tale of how we wrapped our world in plastic. Knowing plastic resides inside us and most other living creatures is deeply disturbing but I am grateful to learn how we got here. *Consumed* explains how slick advertising played a crucial role in the plastics revolution, convincing us to ditch refillable glass bottles, discard washable linen diapers and insist on food wrapped in film. We also fell for the deceitfully reassuring lie of recycling – a tale *Consumed* tells in dramatic detail. The only real antidote to this plastic plague is a truthful tale told well enough to supercharge our desire to do something about it. Consumed achieves this feat admirably'

Simon Clark, co-author of *The Key Man*

'This deeply researched, eminently readable book exposes how consumer products giants deflected one environmental challenge after another to catastrophic effect. Saabira Chaudhuri lays out the hidden history of how the Cokes and P&Gs of the world unleashed lobbyists, lawyers and spinmeisters to evade their responsibilities to the planet with some masterful greenwashing. *Consumed* is the drama-filled, yet highly factual, story of the road to our microplastics crisis'

John Helyar, bestselling co-author of *Barbarians at the Gate*

'Fantastic reporting! This book will entertain you even as it raises your blood pressure – and hopefully it will move you to join the movement to rein in this kind of reckless nonsense!'

Bill McKibben, bestselling author of *The End of Nature*

'Our entire lives are pressed out of plastic these days – but how did that happen? Saabira Chaudhuri's *Consumed* tells the astonishing, bemusing and sometimes enraging story of how we let plastic take over the world. Her reporting transports us back to the nearly plastic-free 1950s, then shows us how we got from then to now. *Consumed* will change how you look at your everyday life, from your garbage bags to your kitchen utensils'

Charles Fishman bestselling author of *The Wal-Mart Effect* and *The Big Thirst*

'*Consumed* is a fascinating and timely book that sheds light on the dirty truth about our use of plastics. Chaudhuri is a deft storyteller, a thorough researcher and a tenacious reporter. Here, she proves to be a compelling and convincing guide to a critical corner of the global environment. Her book should be read by policymakers, corporate titans and anyone with an interest in the future health of our planet'

Duncan Mavin, author of *Pyramid of Lies* and *Meltdown*

'An honest and eye-opening account of how plastic has shaped our world, for better and worse. Essential reading for anyone who cares about sustainability and the future of our planet'

James Piper, host of *Talking Rubbish*

'This book will be essential reading for those interested in how we got into the plastic mess we are now in, as activists seek to hold firms accountable for pollution they generated for decades. From McDonald's hamburger packages to disposable diapers to Coca-Cola bottles, shampoo containers, straws and more, this is a comprehensive look at a profligate plastic past that is now due for a reckoning'

Bart Elmore, author of *Citizen Coke* and *Seed Money*

CONSUMED

CONSUMED

CONSUMED

SAABIRA CHAUDHURI

How Big Brands Got Us Hooked on Plastic

bl!nk

bl!nk

First published in the UK in 2025 by Blink Publishing
An imprint of Bonnier Books UK
5th Floor, HYLO, 105 Bunhill Row,
London, EC1Y 8LZ

A CIP catalogue record for this book is available from the British Library.

Hardback ISBN: 978-1-7851-2032-9
Trade paperback ISBN: 978-1-7851-2216-3
eBook ISBN: 978-1-7851-2187-6
Audio ISBN: 978-1-7851-2188-3

1 3 5 7 9 10 8 6 4 2

Typeset by IDSUK (Data Connection) Ltd
Printed and bound in Great Britain by Clays Ltd, Elcograf S.p.A

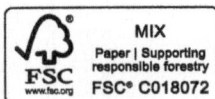

MIX
Paper | Supporting
responsible forestry
FSC
www.fsc.org
FSC® C018072

The authorised representative in the EEA is Bonnier Books
UK (Ireland) Limited.
Registered office address: Floor 3, Block 3, Miesian Plaza,
Dublin 2, D02 Y754, Ireland
compliance@bonnierbooks.ie
www.bonnierbooks.co.uk

For William, Rafi, Zara and Suzy

CONTENTS

PREFACE

In the autumn of 2018, I began working on a story for the *Wall Street Journal* about how bottled water makers were trying to quell concerns many Americans were having about single-use plastic.

To figure out how integral plastic bottles had been to the mainstream commercialisation of water, a drink freely and safely available out of a tap in many parts of the world, I started digging into the past.

Along the way, I noticed that many of the documents and articles I was finding from the 1990s onwards sounded very familiar. Several of the companies currently promising to make their bottles from recycled plastic had made the same promises before, sometimes several times, over the past few decades. Starting in 1990, I counted at least eight instances when Coca-Cola had set a sustainability target for its plastic bottles, usually around recycling, and then failed to meet this. These failures didn't deter Coke from continuing to publish flashy new press releases, offering no explanation about its missed targets while touting new ones.

Given that polyethylene terephthalate (PET), the plastic used to make drinks bottles, is the easiest to recycle of all plastics, why, I wondered, had companies repeatedly set such targets only to miss them? Why was the PET bottle recycling rate in the US below 30 per cent, more than half a century after Coca-Cola and Pepsi had first started putting their soft drinks in plastic bottles?

That bottled water story, eventually published on the *Journal*'s front page in December 2018, was the start of my interest in the deep dependence that consumer goods companies have on single-use plastic.

The vast majority of plastics today are derived from fossil fuels. Plastic production, if it continues on its current growth trajectory, is estimated to account for 20 per cent of the world's oil consumption by 2050. Many plastics are made using a bewildering array of additives that can leach into food and drinks. Scientists are increasingly concerned that such chemicals, even in very low doses, are impacting our fertility, growth and other functions. Tiny plastic particles have been found in everything from foetuses to the human brain, with potentially huge knock-on effects for our health – as well as for biodiversity and the ocean's ability to act as a carbon sink.

Globally, the equivalent of one garbage truck of plastic waste ends up in the ocean every minute. Plastic comprises between 70 and 80 per cent of the litter that ends up in land and marine environments. Most plastics are uneconomical to recycle; more than three decades after US municipalities began including plastics in their kerbside recycling programmes, the country's recycling rate for discarded plastic packaging stands at 13.6 per cent. Globally, just 9 per cent of all plastic waste ever created has been recycled.

Many of the consumer goods companies and supermarket chains I spoke to seemed desperate to distance themselves from the problems associated with single-use plastic. But the more I looked, the more I saw a glaring say-do gap.

In London, where I live, fast-food chain Leon was putting drinks in cups it touted as compostable, without telling customers there was no composting infrastructure available and it was sending these to incinerators.[1] Unilever and Kimberly-Clark were telling people to put biodegradable face and baby wipes in their bins, claiming these would ultimately break down in landfills.[2] Guaranteeing anything will break down in a landfill is a fool's errand – researchers have found 20-year-old corn cobs with all of their kernels intact in landfills, as I'll tell you about in this book. If the wipes did break down, they'd release methane, a potent greenhouse gas.

Sainsbury's announced in 2019 that it had removed plastic packaging from its single cauliflowers, yet whenever I visited the supermarket chain's stores, I saw cauliflowers in plastic. Iceland, the frozen foods chain, said it would eliminate plastic packaging for its own-label products by 2023. Not only does it continue to use large amounts of plastic, but it hands over its online orders in enormous disposable plastic bags instead of the reusable crates other supermarkets use.

Every few months, an email landed in my inbox detailing how plastic waste accepted for recycling was found to have been landfilled, dumped or burned.

At the end of a reporting trip to the packaging R&D centre of an enormous consumer goods company I watched as the company's packaging head took off his (unrecyclable) nylon hairnet and tossed it into a recycling bin without a second thought. Given we had just spent the day discussing how the company was not just trying to improve the recyclability of its products but also change consumer behaviour when it came to recycling, this thoughtlessness when it came to his own discards struck me as absurdly funny but also rather sad.

---

[1] In 2019, Leon switched to using plastic-lined paper cups. These can only be recycled if collected separately since they must go to specialist facilities.

[2] In 2022, Kimberly-Clark switched to labelling its wipes as '0% Plastic', a decision it said was made after 'acknowledging that the rate of biodegradation may vary depending on environmental conditions'.

The idea for *Consumed* was born out of a desire to understand how it came to this. What did the world look like before so many of our everyday purchases came in disposable packaging? Why have so many attempts to rein in plastic waste failed? Why have none of the multi-million-dollar solutions bankrolled by many of the smartest, richest companies in the world made a difference? Why is plastics recycling so broken? Are companies trying to mislead us or is something else going on? *Consumed* investigates how, in the years after the Second World War, plastics makers latched onto disposability as a business model, convincing brand owners to pivot away from reusables (or from no packaging at all) to throwaway plastic. Companies like Coca-Cola were initially reluctant to shift to disposables. But when they did, they found disposability so profitable that they doubled down on it.

Notably, consumers were reluctant to shift too. Contrary to what companies would have us believe, regular people weren't clamouring for disposable diapers, coffee cups, soda bottles, shampoo sachets and food packaging. Companies poured millions of dollars into convincing us to shift to disposables. This didn't just change how business is done, it also fundamentally altered the way we live. Think of toddlers kept in disposable diapers for far longer than their parents ever wore cloth, our love affair with bottled water and our seemingly insatiable appetite for snacks and coffee on the go. While at first it was us – mankind – who shaped plastics, somewhere along the way plastics took over and began shaping us.

Belatedly, companies woke up to how the business models they relied on, and the consumer behaviours they themselves had created, were presenting an enormous solid-waste problem. But, by then, the companies were trapped in a cage of their own making, unable to spring it without destroying the golden goose within. And so they made overblown claims and offered us band-aid 'solutions' designed so they could continue with business models that allowed them to shirk responsibility for the choices they made. For reasons that will become clear as you read this book, these solutions were doomed to fail from the get-go.

There is, however, another way. The point of investigating how we got here is to arm us all to make better decisions about where we go next. Without understanding this history, there's a real danger that we keep relying on the same failed solutions, using the same flawed assumptions and working within the same unviable system – and that we run out of time to change course.

The takeaways from *Consumed* are valid globally, even though much of the book is based in the US, the largest per-capita creator of plastic waste of any major economy. How the US fares has global implications – often the products and habits that start there make their way to the far reaches of the globe. America has also histori-cally been a large exporter of its used plastic to countries that have a poor track record of managing waste.

As I did the reporting for this book, my own perceptions about the origins of our waste and litter problems, and the solutions we should pursue, shifted. I went from being narrowly focused on plas-tic as a material to seeing our problems as rooted in the overarching convenience model predicated on disposability that plastic has so cheaply enabled and accelerated.

Disposable products and packaging existed long before plastics became mainstream. But the huge functional benefits and many conveniences that plastics offer dirt cheap have rapidly accelerated our use of disposables and created a culture of reckless overcon-sumption, while turbocharging profits for companies.

Convenience and choice offer benefits to human beings, but come with largely unacknowledged, uncounted environmental and health costs that impact us all. Whereas previously one paid more for higher performance and convenience, at some point it seems we fell through the looking glass into a world where it's cheaper to choose the most convenient – and invariably more polluting – option. This mismatch between our now deep-seated expectation of virtually untrammelled convenience and choice on the one hand and our unwillingness to pay for it – or even acknowledge its cost – on the other, is at the heart of the crisis we now find ourselves in.

I'm not anti-plastic and I recognise the many ways in which plastics have improved our lives. Critics who call for a wholesale ban on plastics rarely acknowledge how environmentally damaging alternatives can be.

But plastic's many benefits don't change the fact that we're all massively overusing it. We need a fundamental reset to sharply limit the various types of plastics out there, the chemicals used to make them, and the uses they're approved for. We need a total revamp of how products are priced to account for their impacts on the environment and human health, incentivising – and, in many cases, *requiring* – companies, and indeed all of us, to make better choices.

Big brands have harnessed plastic to shape our everyday habits and expectations in ways that no other physical material has – or likely ever can. Any serious attempt to undo some of the damage that's become so intrinsically tied to using plastic will involve hard choices and big trade-offs. It won't be an easy ride. But it is doable.

PROLOGUE: THE WANDERING GARBAGE BARGE

'This junk can't be erased'

The garbage on board the *Mobro 4000* belonged to a hapless entrepreneur from Bay Minette, Alabama, hoping to make a quick buck.

Lowell Harrelson had heard about how the methane released by garbage could be turned into energy. If he could only get his hands on a large amount of other people's discards, he figured he'd make a killing.

Fifty-two-year-old Harrelson wore large hats, guzzled whiskey, smoked three packs of cigarettes a day and had a vocabulary that would have made a sailor blush. The son of potato and corn farmers, he had no childhood memories of the family ever producing any garbage – or even owning a trash can. He left home to study engineering in California in the early 1950s, returning to work as a builder and plumbing contractor. He later moved to Houston, making enough to buy a 40-acre property complete with a swimming pool, tennis court, three-hole golf course and lake filled with catfish and bass.

But lately his debts had soared and Harrelson had seized upon garbage as his golden ticket back to the good life. The year was 1987 and, for anyone in search of vast quantities of garbage, New York was the place to go. Harrelson, a consummate networker, made a few calls. It didn't take long for one of his contacts to suggest that his best bet was a guy called Salvatore Avellino.

'Sal' wasn't your average garbage collector. He was a Mafia boss – the leader of the Lucchese crime family – who tightly controlled garbage collection on Long Island and didn't take kindly to competition. But Sal didn't come across as your average mobster either. He was a versatile individual, as at ease with Wall Street bankers as he was with the garbagemen on his payroll. A few years later, his notoriety would soar after he was convicted of the murder of two rival Long Island garbage haulers. At the time, though, Harrelson insisted he didn't know Sal was a Mafia boss. He claimed it was only when the FBI showed up on his doorstep demanding to know why he was meeting with the mobster that he realised who he was dealing with.

New York was short on landfill space and Sal was paying to truck the trash his guys collected to dumps in other states. Harrelson ginned up an offer too good to refuse: he'd take the trash off Sal's hands for a lower price than trucking it, no strings attached. Over dinner, he explained to the mobster how he'd make up the money by hiring a boat to get the trash down to North Carolina. There he'd found six farmers with fallow land willing to bury the trash on their farms in exchange for a split of the profits when Harrelson extracted methane gas from it.

The two men, an unlikely pairing with the exception of their mutual enthusiasm for making money, shook hands. Harrelson had his garbage.

He rented the *Mobro 4000*, and the trash – all 3,100 tons of it – was compacted and loaded on board. On 22 March, 1987, the orange barge, towed by a tugboat named *Break of Dawn*, left the Long Island suburb of Islip in Suffolk County.

Its moustachioed captain, Duffy St Pierre, set sail for a privately owned port near Morehead City in North Carolina. Harrelson hopped into a Cadillac to begin the long drive south to meet his cargo when it docked.

He had no idea that what was meant to be a quick journey for the *Mobro* would turn into a months-long quixotic quest followed closely by much of the nation.

When the *Mobro* docked in North Carolina, the stench drew complaints and local environmental officials promptly obtained a court order to stop the trash being unloaded. 'We have enough garbage of our own,' sniffed a spokesman for the state's waste department, which promptly ordered Harrelson to 'get the boat the hell out of North Carolina waters'.

The barge lingered in the port for 12 days – its crew pelted with rotten eggs by angry residents – before being unceremoniously escorted back to sea by the US Coast Guard. St Pierre then sailed towards Louisiana. This time, the *Mobro* was allowed to rest for just a few hours, tied to a sweet gum tree on the banks of a deserted channel south of New Orleans at the mouth of the Mississippi. Then a court order kicked it back out to sea.

Just as in North Carolina, Louisiana's state officials said they had their own garbage to worry about and didn't need any more. The Southerners particularly did not want New York garbage. 'We already have deep and bitter memories of the Civil War,' said a Louisiana congressman gravely.

Louisiana's governor threatened to deploy the National Guard to shoot at the barge if it returned. He had environmental officials working over the weekend to track down the *Mobro* to make sure it was no longer in state waters – or, worse, dumping its load overboard. The barge was eventually found 40 miles south-west of the sweet gum tree, moored to a Conoco oil platform in the Gulf of Mexico.

Frantically making calls to local government officials from a $29-a-night motel in New Orleans, Harrelson was starting to panic. In his wildest dreams, he had never thought he'd be saddled with a giant barge full of garbage that had no home. Keeping the boat in the water was costing him $6,000 a day in salaries, insurance, fuel and rent. His hopes of getting rich on methane were fast disappearing. He now just needed to cut his losses and find somewhere to dump the garbage.

Much of the trash came from Islip, a town nestled on the south shore of Long Island in Suffolk County, where landfills were reaching capacity. By now, its officials and residents were squirming with embarrassment. 'Islip's garbage has become the laughing stock of the Eastern Seaboard', moaned the town's bearded, bespectacled supervisor Frank Jones.

A few days later, Harrelson's garbage problem was officially an international issue. The barge had sailed towards Mexico's Yucatán Peninsula, eluding a US Coast Guard cutter carrying Environmental Protection Agency officials looking to take samples of its cargo. Mexico said it would not take Long Island's trash either. Indignant officials at the US Embassy in Mexico City demanded a meeting with the EPA to discuss 'the barge that is attacking Mexico'.

One Mexican newspaper darkly wrote that the attempt 'serves to illustrate once again the scorn that certain sectors of US society feel toward Mexico in particular and Latin America in general'. Environmentalists complained that Mexico was considered 'a trash can for the United States'. The garbage barge and its aromatic cargo had become a political football.

The Mexican navy sent out two planes, a helicopter and four ships to keep the barge out of its waters. The US allocated two Coast Guard planes to keep track of the *Mobro*. Belize, which Harrelson tried next, sent out its defence forces to keep the barge at bay.

Enterprising fishing boats had begun charging $100 an hour to ferry reporters to the 96-foot tugboat that pulled the *Mobro* along.

Prologue: The Wandering Garbage Barge

Newspapers carried solicitous profiles of the boat's 57-year-old captain St Pierre and his three-man crew – a first mate, an engineer and a deckhand – describing what they ate for breakfast (eggs, sausages, fried beans and grits or biscuits), what recipes they cooked for dinner (shrimp jambalaya) and what movies they watched (*To Live and Die in LA*, *Witness* and *Cat Ballou*) on video tape to pass the time. After ten weeks on the boat, which was nine more than originally intended, St Pierre's patience was wearing thin. He longed to return home to Luling, a Louisiana town that sat just south of the Mississippi river, where his family – who had long worked in the shrimp business – had lived since before the Civil War. The captain supported Harrelson's methane-producing idea, blaming 'dodos with red tape' for stopping the businessman from fulfilling his dreams. 'It's like Russia,' he told the *Washington Post*.

The story took on a life of its own. St Pierre's mother was interviewed. Harrelson's sister decided to make the most of her beleaguered brother's misfortune, designing $6 commemorative T-shirts showing the barge and tug against the Manhattan skyline with the words 'Sink the Stink' and 'Garbage Tour'. She even hired a garbage truck to use as an advertising backdrop for Harrelson's nephew and nieces to model the shirts in front of.

The family tried to keep Harrelson in good spirits. On his 53rd birthday, a few weeks into the *Mobro*'s voyage, his wife and kids surprised him with a barge-shaped birthday cake covered with smaller cakes modelled after little bundles of trash.

Newsday assigned 25 people to cover the barge's journey. It set them up in a separate room without any windows or clocks. The rest of the newsroom called it 'the Garbage Room'. The publication tracked the *Mobro* throughout its journey. One tenacious reporter later confessed she got so seasick from all the tugboat's rocking that she had to pause her interviews with St Pierre every few minutes to run into the tug's tiny toilet and throw up.

Consumed

A pair of native New Yorkers living in Colorado came up with a song about the barge that proved so delightfully catchy that radio stations started carrying it. The first few lines went:

> I chewed some gum a week ago, and threw it in the can,
>> Taken to the Islip dump, by the garbage man.
> We're all full up, we're all full up, we can't take no more waste,
> What will we do, what will we do? This junk can't be erased.

Chorus:

> The garbage barge, the garbage barge,
> It sails the seven seas.
> No harbour for its funky gore,
> At sea, eternally.

After Florida rejected the *Mobro* too, EPA inspectors wearing white hooded hazmat suits, green gloves, mustard boots and breathing apparatus clambered onto the barge to take stock of its contents. Their findings were anticlimactic: they discovered mostly paper and cardboard, plastic bags, wood, clothes and shoes. There were a few tyres and mattresses on board. A lonely pink yo-yo later emerged.

Then, after 47 days at sea, there was a ray of hope for Harrelson. The owner of Little San Salvador – a 5½-mile-long, 2½-mile-wide island in the Bahamas – said he'd take the trash to use as fill. He'd ask just $100,000 for the favour. But before Harrelson could agree, the Bahamas government got wind of the matter and put its foot down.

In mid-May, by which time the *Mobro* had been out at sea for nearly two months, Islip officials said they would permit operators to increase the capacity of the town's local landfill and rescind a ban on commercial garbage, allowing it to accept trash from the barge. The cost to Harrelson to dump the trash would be $40 a ton, or $124,000 in total.

Prologue: The Wandering Garbage Barge

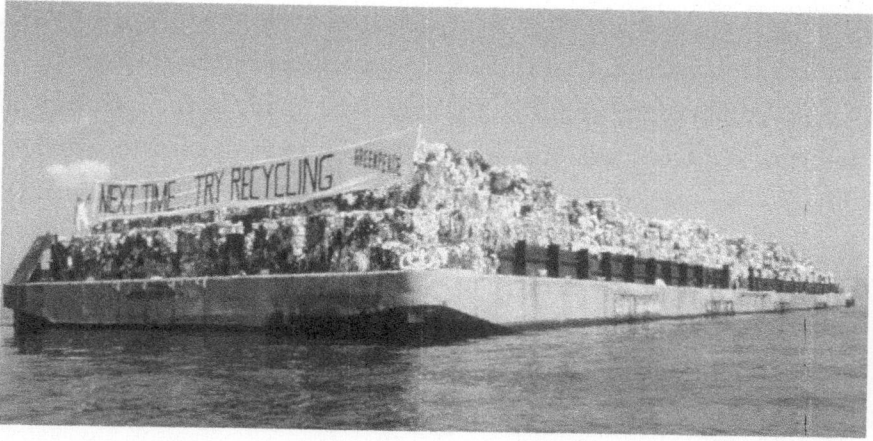

Greenpeace activists hung a banner on the *Mobro* that said 'Next Time Try Recycling'. Image courtesy of Greenpeace

You'd think the end would have been near, but Harrelson's trials weren't over yet. New York City officials objected to the trash being unloaded on health grounds and the *Mobro* was turned away by police boats as it tried to dock, leaving Harrelson wringing his hands in despair. 'I feel like a pilot who has run out of altitude and ideas,' he said upon hearing the news. 'And I'm losing air space quickly.'

Eventually St Pierre turned the barge around and docked at a federal mooring in Gravesend Bay near Brooklyn. The captain himself was finally allowed to go home. But the *Mobro* sat for a further 90 days, becoming a foetid tourist attraction as various officials fought to keep its contents from being unloaded in their boroughs.

New York City agreed that the trash could be burned in Brooklyn, with the ash buried in an Islip landfill. But a Brooklyn judge issued a temporary restraining order, incinerator workers refused to go near the trash, and the fate of the barge became tangled up in court. After a seven-day trial on whether burning the trash in Brooklyn was legal and safe, a judge ruled that the burning could go ahead, under the condition that this be closely supervised by inspectors.

On 2nd September, 1987, the trash was finally burned. By now it had sparked Congressional hearings and hundreds of news articles. The *Mobro* had turned into an international symbol of America's garbage crisis.

The US was producing close to 230 million tons of garbage a year, nearly four times as much as Japan, despite having only twice as many people. Packaging was an increasingly big culprit. Between 1960 and 1987, the amount of packaging in garbage had increased by 80 per cent to make up a third of what Americans threw away.

A study by the Environmental Protection Agency claimed that at least 27 US states would have severe landfill problems within a decade. The Worldwatch Institute, an environmental research group, said that by 1990, roughly half the cities in the US would have run out of landfill space.[1]

A prominent New York assemblyman warned that the garbage problems extended well past Islip. 'Our garbage will not fall off the edge of the earth. It will return to us in the form of polluted air and water and could bury our economy as well – unless we act wisely now,' wrote Maurice Hinchey in an op-ed for *Newsday*. 'The barge saga has made Islip's garbage a bad joke, but the shortage in waste disposal capacity is statewide, even nationwide.'

Amid the hysteria about landfills, plastics took the most heat. 'The very durability of plastic has conferred a curse,' wrote the *New York Times* in September 1987. 'Our indestructible garbage is killing wildlife and threatening to choke human communities on their own waste.'

Frank Jones, the Islip town supervisor, wrote an editorial for a local New York paper. 'America, the richest nation in the world,

---

[1] These various prognostications of a landfill crisis turned out to be inaccurate, but we'll get to that later.

is burying itself in its own garbage,' he warned. 'Some of the most frightening aspects of our refuse flow are those created by plastics, especially the array of packaging materials that are designed, for the most part, to protect the product and attract the buyer.'

Islip, Jones promised, would begin a recycling programme that gave residents containers to separate paper, glass and metal. Notably, plastics were missing from the town's plan. Plastics, wrote Jones, 'should no longer be landfilled, they cannot be recycled, and should not be burned because of the possible pollutants produced. What good is separation of plastics, when there is no environmentally sound way of disposing of them?'

*

The story of the *Mobro 4000* and its unwanted load of trash encapsulates a problem the world is still grappling with today: what do we do with all this stuff – much of it designed to be used once but last hundreds of years – that we keep creating and discarding in ever-increasing quantities?

The *Mobro* sparked a widespread awakening among regular Americans about how much trash they were generating. It also put single-use plastics in the hot seat in a big way. Paper, glass and metal could be recycled and food waste composted. But, at the time, plastics could only be landfilled, burned or littered.

This realisation created a public-relations challenge that threatened to be existential for both the chemical companies making billions from plastics and the consumer goods companies that had built veritable global empires on the back of single-use plastic in the decades after the Second World War. The companies and their trade bodies attacked the threat head-on.

Pampers-owner Procter & Gamble said it would make a compostable diaper. Mobil sold trash bags it said would harmlessly degrade and disappear. McDonald's vowed that its polystyrene foam burger containers would be widely recycled. And Coca-Cola promised large numbers of soda bottles made from recycled plastic, even as it fought laws designed to improve recycling rates. None of these solutions worked out. They did, however, help companies quash anti-plastic sentiment at its height, defeat bills that threatened profits, and win over consumers and lawmakers.

In the years that followed, companies faced with slowing growth in the West began looking east to emerging markets like India. To drum up demand for their brands among people who had long gone without them, they pumped trillions of small, ultra-affordable plastic sachets into places that didn't even have basic waste collection in place, let alone recycling. For the most part, they got away with it; no politician has a strong enough stomach to ban products ostensibly catering to a country's poorest people.

Then, in 2015, a new backlash against single-use plastics broke, quickly going global. It began with the sorry tale of a turtle found off the coast of Costa Rica with a plastic straw lodged in his nose. This backlash is ongoing and the public's concerns aren't limited to worries about landfill capacity. They now encompass plastic's contribution to global warming, the health effects of the endocrine-disrupting chemicals and thousands of other additives found in plastics, and the potentially seismic biodiversity and health impacts of microplastics, which we've only recently learned are absolutely everywhere.

Consumer brands have scrambled to defend their sprawling plastic empires, trotting out many of those same, decades-old failed solutions to win over the public and setting new green targets they've already started to miss. Yet, as our understanding of the full ramifications of our current convenience model deepens, there are signs that – unlike the wave of concern in the 1980s that soared and then crashed out – this backlash could in fact spark sustained positive change.

PART I

PLASTICS UNDER PRESSURE

CHAPTER 1

THROWAWAY LIVING

A Super Highway With No Exit

In November 1963, Lloyd Stouffer, the editor of *Modern Plastics* magazine, addressed hundreds of plastics industry executives who had gathered for a conference at the Sheraton Hotel in Chicago. Stouffer began by recapping a talk he had given back in 1956 in which he had said 'the future of plastics is in the trash can' and explained that, going forward, the industry must focus on disposability to make money.

'What I had said in the talk was that it was time for the plastics industry to stop thinking about "reuse" packages and concentrate on single use,' Stouffer told the room. 'For the package that is used once and thrown away, like a tin can or a paper carton, represents not a one-shot market for a few thousand units, but an everyday recurring market measured by the billions of units.'

Now, just seven years later, a beaming Stouffer congratulated his audience on its success. 'You are filling the trash cans, the rubbish dumps and the incinerators with literally billions of plastics bottles, plastics jugs, plastics tubes, blisters and skin packs, plastics bags and films and sheet packages – and now, even plastics cans,' he exclaimed.

Already, in the 25 years leading up to 1962, the plastics industry had grown eight times faster than the US economy as a whole. The steadily falling price of plastic and the widening cost advantage it offered over other materials meant that plastics' future could only be rosy, maintained Stouffer. 'The happy day has arrived when nobody any longer considers the plastics package too good to throw away,' he told his audience.

*

The plastics industry's roots hark back about 150 years to a man called John Wesley Hyatt, the 28-year-old owner of a print shop in Albany, New York.

A blacksmith's son with a creative streak – he had once patented a knife sharpener – Hyatt had responded to a call for help from the largest billiard ball supplier in the US. Phelan & Collender was offering $10,000 in gold to an 'inventive genius' who could find a substitute for the increasingly scarce ivory it relied on to make its balls. Eventually, Hyatt made billiard balls coated with collodion, a liquid that printers used to protect their hands from scorching temperatures. The balls were highly flammable and had a tendency to make loud gunshot noises upon colliding. Unsurprisingly, that made them less than ideal for billiards and there's no indication Hyatt collected the reward for his sub-par offering to parlour sport. But the process led him to discover something called celluloid, which was effectively the start of the synthetic plastics industry.

Celluloid was formed by dissolving nitrocellulose paste under heat and pressure with camphor, turning it into a tough, whitish mass. The material's consistency changed with the amount of camphor added, allowing for substances ranging from hard and horn-like to soft and rubber-like. Celluloid was a 'thermoplastic', a mouldable mass that held its shape after heat and pressure were withdrawn and which could be reheated and remoulded. It was easy to colour and shape. Notably, unlike fiddly wood and metal, it could be mass-produced using machines.

Celluloid proved a good substitute for materials pulled from all over nature – tortoiseshell, coral, amber and especially ivory – used for hair combs, eyeglass frames, jewellery and hat pins. It held the promise of democratising luxury goods and of easing guilty consciences by sparing elephants and tortoises. 'It will no longer be necessary to ransack the earth in pursuit of substances which are constantly growing scarcer,' proclaimed an 1878 pamphlet from the Celluloid Manufacturing Co., set up by Hyatt and his brother to market the new material.

Celluloid marked the advent of a new age, one in which the quest to find new materials was shifting from the hands of grizzled adventurers who braved the high seas to bespectacled inventors combining chemicals in laboratories. While humanity was previously at the mercy of nature, taking what we could find and devising useful applications for it, the plastics age put humans at the helm, circumventing nature to create materials that held the properties we desired.

Bakelite, another early plastic, also came about through the trials of a lone inventor looking to create man-made materials that would ultimately free humans from the vagaries and limitations of nature. The natural substance in short supply this time was the female bug secretions used to make shellac, an insulator for electrical cables. It took 15,000 beetles about six months to make what was needed for a pound of shellac. This was harvested through scraping the hardened deposits from trees, a labour-intensive and increasingly untenable way to meet surging demand from the electrical industry.

The son of a cobbler and a maid from Ghent in Belgium, Leo Baekeland in 1907 created Bakelite by mixing formaldehyde with phenol – a by-product of coal that was commonly used as a disinfectant – to create a black-brown material. It was the world's first truly synthetic plastic. Bakelite had a high resistance to electricity and heat, and was heavily used not only in the electrical and automotive industries, but also to make telephones, washing machines, jewellery and radio cases.

In those early years, there was rampant confusion about what the word plastic – derived from the Greek *plastikos*, which means capable of being moulded – even meant. For a long time, it was a loose term that referred more to a manufacturing process than a product. In 1925, the industry's first trade journal, *Plastics*, floated a definition, saying plastics were a commercial class of substances 'worked into shape for use by moulding or pressing when in a plastic condition'.

Today, plastics are commonly described as polymers – composed of long chains of repeating molecules called monomers. They're mainly made up of carbon and hydrogen atoms which mostly come from crude oil, natural gas and coal. The polymers are mixed with additives, which give them various properties, such as resilience and flexibility.

Following the advent of Bakelite, other plastics – like rayon, nylon, polyethylene, polypropylene, cellulose acetate, polystyrene, polyvinyl chloride, acrylics, melamine and urea formaldehyde – mushroomed. The centre of gravity shifted from individual inventors to scientists housed at deep-pocketed chemicals makers like E.I. du Pont de Nemours & Co.

DuPont, as it was known, had started out life in 1802 as a saltpetre mill on the humid shores of the Brandywine river in Delaware. It went on to supply the explosives used in five wars, but in 1910 began looking for ways to diversify after the US government started building its own munitions plants.

Starting in 1915, DuPont acquired or licensed cellulose-based materials like celluloid, rayon and cellophane film. It went on to invent nylon, the world's first all-synthetic fibre, which DuPont savvily marketed as made from 'coal, air and water'. Nylon quickly achieved cult status among women. Stockings made from the sheer fibre famously provoked fights in department stores when they launched in May 1940. 'Girl Collapses, Woman Loses Girdle at Nylon Sale' reported the *Johnstown Democrat*.

The first four decades of the 20th century saw these plastics secure a significant foothold in America. DuPont, for the first time, began advertising directly to consumers – and, in particular, housewives – in a big way. Through the 1920s and '30s, DuPont hired writers for women's magazines like *Good Housekeeping*, along with interior decorators, home economists and etiquette experts, to evangelise about its synthetic crush-free fabrics, wipeable furniture and unbreakable household ornaments.

By 1940, the US was producing 276 million pounds of plastic per year, compared with just 5 million pounds 20 years earlier. But it was the Second World War that really kicked plastics into high gear, giving the burgeoning industry the credibility and technological know-how it needed to become a giant.

In December 1941, after Japan bombed Pearl Harbor, the US entered the war, prompting a clampdown on the non-military use of all high-priority materials needed for defence, such as metals, rubber and silk. These materials were also heavily imported and, as trade routes collapsed, they began to run short. It was here, in the crucible of war, that plastics arrayed themselves across the spectrum of human need, securing an integral role that would not – and, in time, could not – be rolled back.

In transportation, plastics were used on the ground, in the sky and on water. DuPont's nylon was now needed for parachutes for soldiers and airmen. Rayon was used alongside rubber to make parts for bombers, combat cars and transport trucks. It retained its strength under the high-friction heat of heavy-duty driving, as well as reducing aircraft weight, allowing for more bombs, higher speed and longer range.

Lucite, a lightweight acrylic plastic also made by DuPont, was used to make salt-water filters for marine engines and transparent enclosures for aircraft and tank windows. Its production increased tenfold during the war. Polyvinyl alcohol, another plastic, was used to line paper, cork, cloth and ceramics, making them stronger and moisture-resistant.

In the realm of food storage, DuPont's cellophane quickly became key to feeding soldiers, allowing ships short on cargo space to carry dehydrated and compressed eggs, milk, meats, vegetables and soups in wrapped moisture-proof packages. Biscuits, bouillon powders, cigarettes and fruits covered in cellophane stayed fresh in all climates.

When it came to equipment and industry, plastic variants meant cellophane and other wraps would also protect rifles, ammunition, airplane fuel tanks and truck parts against rust and mould.

For all of these areas where plastics were proving transformative, there were as many everyday uses they quietly fulfilled for America's troops. Nylon bristles meant soldiers could keep brushing their teeth and hair after supplies of hog-hair bristles, previously imported from China and Russia, were cut off. Contact lenses shaped from plastic proved far more convenient than foggy spectacles. Army doctors used plastic for surgical sutures and catheters, while dentists used it to make plates.

In the span of a few years, plastic became so ubiquitous that a wartime general would have relied on it across every area of his military operations, a remarkable accomplishment for a substance still in its infancy when most of those generals were born. 'The raw materials that went into dishpans, clothing, homes and automobiles now float, fly and roll as units of America's vast engine of war,' proclaimed DuPont in 1942.

During the First World War, the US had relied on other countries for supplies, but this time its manufacturing prowess meant the roles were reversed, crowed DuPont. 'We are a strong nation,' it wrote in its in-house magazine. 'If we still had to rely wholly upon animal skin and horns, upon plant seeds and fibres, upon mineral ores and ashes, our lives would be vastly different and more difficult.'

Of course, it wasn't just DuPont. Other companies like Monsanto Chemical and Dow Chemical also churned out plastics for the war. Monsanto's vinyl acetal was used to make ponchos, hospital sheeting and life jackets. Dow's Saran made insect screens for troops, its Styron made radio and electrical parts, and its Ethocel helped shape metal for airplane manufacturing.

Between the start of the war in 1939 and its end in 1945, the plastics industry's sales jumped 325 per cent. As victory looked assured, the industrialists of America trained their sights on transforming many of the same high-tech discoveries made during the war into products for the country's kitchens, living rooms, nurseries and bathrooms.

In August 1945, four days before the atomic bombing of Hiroshima, DuPont executive J.W. McCoy outlined the huge opportunity to fulfil all the demands pent up by the American public. But McCoy also warned that satisfied people were 'stagnant' people. His answer to this seeming conundrum was to urge merchandisers to satisfy immediate desires but simultaneously work to ensure that 'Americans are never satisfied'.

*

The culture of convenience and disposability that was in full swing by the time the *Mobro* set sail from Long Island in 1987 had been simmering since the turn of the century. But it really took off in the

1950s, encouraged by marketing campaigns that lauded the many benefits of throwaway living.

Through the Second World War, with many materials in short supply, recycling and resource conservation were a way to show one's patriotism. Americans buying a new tube of toothpaste had to hand in an old one in exchange. Utilities companies launched a campaign to teach housewives how to prolong the lives of their refrigerators, kettles and other appliances through careful use. They also paired monthly electricity bills with booklets urging Americans to cut and reconnect damaged cords and repair toasters. For more complex jobs, companies ran free training sessions.

'Buy only what you NEED' urged an ad from the US War Advertising Council. 'And before you buy anything remember that patriotic little jingle: Use it up. Wear it out. Make it do or do without.'

Britain went further, curtailing both products and packaging: shoppers took home soap with no wrappers, toothpaste without cartons, and cigarettes in paper packages with no inner foil or outer cellophane wrappers. Housewives suddenly found that products came in one size only. Envelopes were reused several times – new addresses were simply written on stickers pasted over the old ones.

After the war wound down, though, boosting consumption and supporting businesses at home was seen as the patriotic path to restore countries' economic health. And what better way to do so than to buy stuff designed to be used once and then thrown away?

An August 1955 article in *Life* magazine encapsulated the sensibility of the day, showing a family gleefully throwing an array of disposable plates, napkins, cups and straws into the air beside an overflowing trash can. 'The objects flying through the air in this picture would take 40 hours to clean, except that no housewife need bother,' it enthused. 'They are all meant to be thrown away after use.'

If cleaning something was messy – think washing out the dog's china dinner bowl – or required extra care – the baby's glass bottle – the industry's answer was to develop a disposable plastic version. DuPont mined the habits of its own workforce for consumer insights and enthusiastically concluded that housewives 'have found disposables a great aid in saving time, making work easier and guaranteeing cleanliness'.

But within that same report extolling the wonders of disposability in the micro setting of the household was a line portending the macro implications of this new trend: 'DuPont housewives are buying bigger trash cans, for each year they use more and more disposables.'

*

The Second World War had seen the largest influx of women into the American workforce on record. While many stopped working initially after men returned from the frontlines, a war-born desire for convenient weeknight meals, easy-to-clean surfaces and wrinkle-free clothes did not disappear.

'Plastics are here to free you from drudgery', *House Beautiful* explained to its readers in 1947. 'Already – and they are still in their adolescence – they are giving millions of women the priceless gifts of more leisure and more energy.'

The magazine was not looking to be hyperbolic when it told readers that, for them, plastics ranked alongside the combustion engine and the telephone in revolutionary importance. 'They are proving themselves on the testing ground you know best – the housekeeping arena where the traditional battle with dust, grime, stains, scratches and shabbiness has made the housewife's profession synonymous with slavery,' it said. 'They are responsible for the three greatest little words in the history of house cleaning: damp-cloth cleaning.'[1]

The baby boom, which began in 1948 and lasted for most of the next two decades, meant many women soon had growing families to cook, wash and iron for. In the decade to 1956, 10 million US homes were built, dwarfing any previous home-building programme. Many were larger suburban dwellings that needed more upkeep.

Through the 1950s, women re-entered the workplace in droves; by 1959, the number of married women holding down jobs outside the home was double that in pre-war America. Dual incomes helped a booming post-war economy, translating into more spending power and a strong desire for leisure time. Plastics makers seized the opportunity to sell time-strapped women surfaces that repelled dirt or could be wiped down, unbreakable dinnerware, and scratch-proof furniture that withstood the clumsiness of children.

Growing incomes fuelled a desire to move up in life. 'If basic human needs are few, civilised human desires are infinite,' DuPont exclaimed in its employee magazine in 1960. It buoyantly described how American households had broken away from an old tradition of living in the same house and preserving the same furniture for generations and were now regularly upgrading to more spacious dwellings which inevitably required buying new furniture.

There was also a new emphasis on hygiene. 'In the old days people could see nearly all the food they bought,' began a DuPont magazine ad in 1942. 'There it was, open to the dusty air in barrels and boxes, handled by almost everyone, exposed to dirt and germs. Freshness quickly vanished – flavour too. Be thankful you don't

---

[1] It wouldn't become apparent until decades later that such easy-to-wipe surfaces made from vinyl often contained phthalates, a class of chemicals that have been linked to reproductive, neurological and developmental issues.

have to shop in those old-fashioned unsanitary stores. That was B.C. (Before Cellophane).'

Cellophane was invented and patented in 1908 by Jacques Edwin Brandenberger, a Swiss textile chemist. Made from wood pulp, in its true form cellophane is biodegradable, although additives and coatings used in the production process may not be. If left in a landfill, cellophane can remain intact for decades.

When DuPont bought the American rights to cellophane and started making it at a plant in Buffalo in the spring of 1924, it cost $2.65 a pound. Some customers considered it so precious that they stored it in a safe. At first, only expensive chocolate and upscale perfume was packed in cellophane. Over the next two decades, varieties of cellophane mushroomed and its average cost plummeted to just 33 cents a pound, allowing DuPont to push it as the plastic wrap that could be used for just about anything.

Cellophane-wrapped pre-cut meat and vegetables helped fuel the rise of the modern self-service grocery store. Items previously weighed, freshly cut and sold at butchers, greengrocers and fishmongers were now sold under one roof. DuPont's data showed that cellophane spurred consumption. Plastic-wrapped apples glistened on shelves, while marbled cuts of steak looked pink and fresh for days, which DuPont said 'stimulates the buying impulse'. Buying from such modern shops was quick and easy, but shiny packaging also added a touch of glamour to the experience. 'A new atmosphere of congeniality has been created,' wrote DuPont. 'Shopping for food has become fun. Men who would rather have faced a firing squad than have done the family marketing have become avid boosters of the super-markets.'

DuPont invested heavily in an army of smartly dressed cellophane salesmen, who made door-to-door calls. The men asked housewives what they looked for when they shopped and how they baked, transferring those insights to the 12,000 companies across 30 different industries that made up DuPont's cellophane customers. Among the many promising opportunities DuPont saw was in wrapping cakes and bread with cellophane, which it promised was twice as moisture-proof as wax paper.

DuPont showed companies how they could use cellophane to sell multiple items at once, tempting consumers to make bigger purchases. 'Making more money for the customer – and thus for DuPont – is the central thread in cellophane's sales history,' explained the company.

Cellophane also allowed companies to sell single servings, like half-slices of bacon designed for small pans, and individually wrapped single-serve portions of coffee that 'save precious time in the kitchen for the housewife'. These were a prelude to the far greater numbers of single-serve portions of everything from nuts to fruit sold at grocery stores and casual-dining chains today.

There were early signs of the havoc such ubiquity would wreak on the environment in a single line buried in a 1941 DuPont article dedicated to celebrating cellophane. 'Birds even use it in building their nests,' the company enthused. At the time, plastics were still seen as environmentally benign, and birds' use of them as cute.

To win over retailers who might be reluctant to adopt the plastic wrap, DuPont brought in surveys indicating that cellophane kept food fresh for longer and stopped housewives from patting vegetables and pinching fruit. It also stopped more discerning customers from cherry-picking the best produce and leaving behind sorry-looking stragglers a store couldn't sell. 'Cleanliness and Cellophane are synonymous,' said DuPont, explaining that unpackaged food was prone to spreading disease.[2]

---

2 Of course, this wasn't necessarily true – plenty of fresh fruit and vegetables were sold unwrapped without such side effects. Many years later, a growing number of studies would show that chemicals and tiny plastic particles can leach from packaging into food and that bagging salad can encourage the growth of salmonella.

Du Pont and other plastics makers poured money into advertisements that touted how plastic could be used to sell single-serve portions of everything from coffee to bacon. © *Saturday Evening Post*

A few years after the war ended, polyethylene, another plastic wrap, began appearing on shelves. While cellophane was great for light produce like spinach, robust polyethylene proved ideal for carrots, radishes, oranges, potatoes and onions. Wrapping produce cost more than selling it loose, but companies more than made up for this as demand rose and shipping costs for items like root vegetables – now packed minus their leafy tops – declined. Fruit and vegetables, once sourced locally according to what was available each season, were increasingly wrapped and transported for consumption all year round.

Fresh food producers embraced transparent packaging wholeheartedly because it allowed them to slap their brand names on the plastic. 'It's just a matter of time before all our fresh vegetables will be packaged,' said the general manager of F.H. Vahlsing, one of the largest fresh vegetable shippers in the US, in 1954. The company was so bullish on plastic wrap that it took the unusual step of opening its own polyethylene bagging plant in Texas.

Plastic wrap also slashed costs for retailers by eliminating the need for queuing and counter service. It cost just 2 cents per sales dollar for shops to handle topped carrots in bags, compared with 14 cents per sales dollar for clerks to weigh and price carrots in loose bundles that had their tops still on. By 1959, 90 per cent of fresh carrots sold in the US were packaged, up from just 1 per cent at the start of the decade.

A less obvious benefit of pre-wrapping fruit and vegetables was that they didn't end up on the floor where people could slip on them. A Los Angeles supermarket chain reported a sharp drop in insurance costs in 1959, telling the *Wall Street Journal* that, before it began putting grapes in packages, 'we lost much more on lawsuits than we made on the grapes'.

Meat packed in polyethylene could go straight into the home freezer, since the plastic remained flexible at temperatures well below zero. By the 1950s, the freezer was the fastest-growing home appliance in the US behind the television. The two went hand in hand. The more families watched TV, the less time they wanted to spend in the kitchen cooking. Accordingly, demand soared for plastic-packed, frozen pre-fried chicken, pies, peas and ready meals that just needed to be popped into the oven or a pan of boiling water. Clarence Birdseye, who pioneered the modern frozen food industry, was an early adopter of cellophane wrapping for his products. One astute journalist noted that the astronomical growth rate of frozen foods exactly matched that of plastics.

Pre-wrapping hastened the demise of the old counter-service stores. While in 1933 there were around 300 self-service grocery stores across the US, the number jumped to 85,000 in 15 years. They offered lower prices and soon two-thirds of all food purchases were made in them.

Beyond the kitchen and supermarket, one of the most notable shifts linked to plastic packaging was the amount of coffee Americans began consuming, much of it on the go. Coffee breaks among office and factory staff caught on soon after the war ended. Employers disapproved of staff wandering out to restaurants on company time, but there was no stopping workers from seeking their mid-morning caffeine fix. 'Whether management likes it or not, morning coffee is in the office to stay and it appears that the custom is on the increase,' wrote the Paper Cup and Container Institute in 1950.

Paper cups had been around since the early 1900s, developed to curb diseases spread by shared metal cups at public water fountains, but they were limited to cold drinks. Wax linings melted in contact with hot drinks, while thicker, unlined paper imparted a papery taste and smell that drinkers disliked. Soon after the Second World War, the problem was solved by a company called Lily-Tulip Cup Corp., which developed a plastic-lined paper cup. 'Eliminates dishwashing completely,' underscored a Lily-Tulip ad. 'Has the look and feel of fine china and because of its plastic coating it's the paper cup you never taste!' Lily-Tulip also developed a snap-over lid for the cup. Both were embraced by consumers.

To keep staff on site, employers began installing coffee vending machines equipped with disposable plastic-lined cups. A nickel got you a 6-ounce hot drip coffee. The cups quickly increased coffee sales. The drink began appearing in hotel rooms, schools, roadside stands and hospitals, as well as on trains, planes and even buses, many of which added snack bars on board. Employers who previously gave their factory workers drinks in reusable cups switched to the disposable ones.

By the first half of the 1950s, *Barron's* magazine was reporting that hot coffee had become the biggest-selling beverage in the US, its volume exceeding that of all others combined. Americans were using and tossing away 3.4 billion cups a year from vending machines alone. The cups also caught on among soft drink makers, who used them for fountain drinks.

In the decade that followed, DuPont, Monsanto and Standard Oil poured millions into experimenting with various plastics for drinks bottles. Their efforts paid off. Coca-Cola and Pepsi both launched their soda in plastic in 1970, accelerating the demise of returnable glass bottles. The light, portable plastic bottles allowed them to sell ever-larger quantities of sugary soda. In offices and canteens, glass jugs for water coolers were steadily replaced by plastic.[3]

As disposable packaging paved the way for Americans to drink more frequently, they ate more often too. The nation's appetite spread beyond the traditional three meals a day and into a state of seemingly constant indulgence, helped by packaged food that appeared just about everywhere. Plastic itself had become a driver of appetites once constrained by time, place, expense or just convention.

---

[3] Many years later, single-serve plastic bottles would turn bottled water into the largest drink in the US.

Pre-wrapped pretzels, donuts, sandwiches, cookies and potato chips began popping up in vending machines, convenience stores, supermarkets, roadside stands and gas stations. Once packaged in waxed glassine paper,[4] by the 1970s these products came in plastic which was easily adapted to high-speed machinery, allowing manufacturers to slash costs. The new packages – increasingly made from multiple kinds of plastic bonded together with paper and aluminium – also meant products could now sit in crates and on shelves for months, allowing companies to expand distribution.

Cookies and chocolates proved particularly popular, with Americans' consumption of sweet goods soaring 65 per cent in the decade to 1968. Many of the new packaged foods appearing on shelves were largely devoid of nutritional value. Plastic-wrapped coffee creamers contained no milk or cream, while dried beef stroganoff contained no beef.

In the decades to come, such foods – high in salt, sugar and fat – would be classed as 'ultra-processed', but even in the early 1970s nutritionists were warning of their ill effects. In 1971, the *Wall Street Journal* reported that 25 per cent of the American population was considered 'seriously overweight'. The paper also reported that 98 per cent of Americans had one or more decayed teeth – you'd hope this was among adults only.

Competition between brands intensified and a package's potential to act as a travelling advertisement became just as important as its ability to protect the contents inside. Shiny plastic packages that were easy to print on invited lurid colouring and inventive graphics, giving companies an eye-catching new tool to target young children and teenagers in particular.

In the 1960s, President Dwight Eisenhower's transformative highway initiative, the rise of corporate-permitted vacation and road trip culture, and the commuting spurred by suburban sprawl all

---

[4] Waxed paper was not recyclable and the wax was usually derived from petroleum.

culminated in a big expansion of roadside diners and quick-service restaurants. Their limited menu of hot dogs, milkshakes, burgers and fries allowed a whole family to eat for the price of a single person at a more traditional sit-down restaurant.

Especially popular were drive-ins and drive-thrus – young parents who ate in the car with their kids didn't need to hire a babysitter. Demographic changes helped too: America's teenage population jumped 50 per cent between 1948 and 1962 and youngsters on the cusp of adulthood loved these new affordable spots and the sexy rush of autonomy they conferred. By the mid-1960s, a whopping 70 per cent of America's restaurants were quick-service. Disposable containers and cutlery were key to keeping prices low. Labour shortages and rising wages meant collecting, washing and drying chinaware was increasingly unappealing.

The shift to disposables across the US was perhaps most apparent to state highway cleaning departments, which were suddenly confronted with a big spike in litter. In 1966, the tab for cleaning California's highways was $2.9 million. Two years later, it had jumped to $4 million.

To service the rush of demand for plastic products, a host of new entrepreneurs popped up. By the early 1960s, some 5,000 firms were turning raw plastic resin into finished products. There were about 30 different families of plastics, ranging from vinyls to acrylics. Plastic production stood at close to 6 billion pounds per year, up from just 130 million pounds before the Second World War. 'The plastics industry appears to us to offer almost unlimited markets in the years ahead,' enthused the vice president of Humble Oil.

Just as the price of cellophane had plummeted, the price of other plastics too was dropping: polystyrene, for instance, cost 18 cents a pound by 1961, down 33 per cent from just three years earlier. Polyethylene's price dropped 50 per cent between 1947 and 1962 as more women relied on pre-wrapped, pre-chopped foods, fuelling a further demand boom.

'Chronologically speaking plastics are only a baby,' one fabricator told *Barron's* in June 1960. 'As adolescents they should kick up quite a fuss.'

*

Lloyd Stouffer's seeming naivety – or perhaps wilful ignorance – about the environmental havoc of increasingly dumped and littered plastics was largely reflective of the industry's approach at the time.

The year before Stouffer's address, in 1962, Rachel Carson, an American marine biologist, had published *Silent Spring*, a seminal work describing how overused pesticides were killing millions of birds and harming humans. Carson's book showed how seemingly minor human actions can have enormous, permanent ramifications. It became a bestseller and is widely seen as the start of the modern, mainstream environmental movement. 'Only within the moments of time represented by the present century has one species – man – acquired significant power to alter the nature of his world,' wrote Carson. 'The most alarming of all man's assaults upon the environment is the contamination of air, earth, rivers, and sea with dangerous and even lethal materials.'

Three decades after *Silent Spring*, another book was published – *Our Stolen Future*. Written by Theo Colborn, John Peterson Myers and Dianne Dumanoski, it laid bare in grim detail how tiny amounts of synthetic chemicals were damaging the ability of animals, and likely human beings, to reproduce. In the years that followed, as the world's fertility rate declined and cancer among young people rose inexplicably, more scientists would turn to studying the effects of such chemicals, which they believed disrupted the hormones necessary for normal growth and reproduction and leached from everyday plastic containers into our food and drink.

Back in the 1960s, when Stouffer addressed his audience, the US was generating around half of the world's pollution and pressure was starting to build on the big emitters. 'Our nation's attitude about pollution has changed from apathy to anxiety,' DuPont's vice president Samuel Lenher told a conference in 1968. DuPont invested in air pollution controls for its production facilities. But the prospect of limiting the growth of plastics – or any of its other products – or even seriously looking for an end-of-life solution beyond landfills and incinerators was not on the table.

Through the 1960s, a bigger and bigger slice of plastic was being used to make packaging. In 1966, 20 per cent of all plastic being produced was used for packaging. That percentage was remarkable because it had doubled in just six years. By the decade's end, the percentage of plastic used for packaging had increased further to 25 per cent.[5] The growth in packaging was faster than the increase in the US population, the rise in disposable income and even the increase in the nation's expenditures on consumer goods.

In September 1969, 250 engineers, technologists, economists, ecologists, marketers and package designers gathered in San Francisco to attend the US's first national conference on packaging waste. Much like Stouffer's talk in Chicago, the conference focused on disposability. However, it was organised by the Bureau of Solid Waste Management, a government body that eventually morphed into the US Environmental Protection Agency, and its vantage point was decidedly different. Attendees were asked to examine the problems created by disposable packages which, as one participant

---

[5] Today, 39.1 per cent of plastics globally are used to make packaging, according to the European Environment Agency.

explained, 'are increasingly made up of material which won't burn, break, crush, degrade or dissolve'.

A plastics industry representative who addressed the conference was frank in his assessment of the future: plastics used for packaging would again double in the next seven years as hospitals, airlines, universities, restaurants and industrial cafeterias all switched to disposables. The representative, Thomas B. Becnel from Dow Chemical, the maker of Saran wrap and Styrofoam, acknowledged the waste disposal issues that came with plastics' success. 'It is ironic that the very molecular structure that has made them so popular also creates certain disposal problems,' he said.

Becnel walked the attendees through the various disposal options as he saw them, including landfills, compacting trash and even burying plastic at sea. He dismissed the latter for being too expensive and potentially risky, given that the buried plastic, if not packaged properly, could float to the surface. 'You would have an ironic situation where you would be packaging your packaging waste,' said Becnel wryly, seemingly unaware that plastic trash bags would soon be all the rage.

In his view, the best solution was incineration. His employer Dow, Becnel told his audience, was already burning scraps from more than 500 chemical processing plants at an incinerator in Midland, Michigan, the site of its headquarters. 'We burn polystyrene foam, we burn polyvinyl chloride, we burn a variety of other plastics,' he disclosed. 'And we do this without polluting the air or the small river on which the plants are located.'

The plastics industry's enthusiasm for burning would meet huge pushback in the years to come from communities that didn't want to be anywhere near incinerators. Contrary to Becnel's claims, incinerators burning polyvinyl chloride (PVC), a common plastic used to make everything from shower curtains to vinyl records, were found to produce highly toxic dioxins. The dioxins travelled widely, landing on grass and making their way into the bodies of cows and other animals. Eventually, they entered the bodies of humans who consumed milk and beef. The ash, laden with heavy metals and other toxins, could also leach into groundwater. Burning plastic also yielded relatively little energy and meant new plastics had to be made from fossil fuels.

Perhaps the most notable critic of incineration was Barry Commoner, a Harvard-educated conservationist who documented the effects of radioactive fallout and who, on the first Earth Day in 1970, appeared on the cover of *Time* magazine. Commoner later ran unsuccessfully for the US presidency and went on to become known as one of the founding fathers of the modern ecology movement.

Commoner wore thick-rimmed black glasses and had heavy black eyebrows like caterpillars, set off against a shock of thick white hair. He was deeply concerned by plastics, which he saw as being destined to pollute the earth and a prime example of how man had 'broken out of the circle of life, converting its endless cycles into man-made linear events'. Ecologically, plastics are 'literally indestructible', he wrote in his 1971 book, *The Closing Circle*. 'Hence every bit of synthetic fibre or polymer that has been produced on the earth is either destroyed by burning – and thereby pollutes the air – or accumulates as rubbish.'

Commoner headed the Queens College Center for the Biology of Natural Systems in New York, which produced studies in the 1980s predicting a spike in lung cancer cases, should the city go ahead with a plan to build new incinerators. It never did. At hearings that followed the *Mobro*'s voyage, Commoner testified that, as a cure for America's garbage problem, incinerators were far worse than the disease.

Becnel of Dow, however, saw burning plastic waste as the industry's best bet to tackle waste. 'Incineration represents the ultimate acceptable alternative,' he told his audience at the conference.

Notably, Becnel made no mention of recycling in his speech. By the time the San Francisco conference took place in 1969, there were a befuddling number of plastics on the market and the complexity of packages was soaring, making sorting far too expensive.

As consumption boomed and discards rocketed, some in the industry began to realise that they had painted themselves into a corner from which there was no discernible escape. A packaging designer at the conference noted that the new packaging – often made from combinations of materials that could not be recycled – was a force to 'self-sell', giving products that used it an edge. But within his speech he delivered a warning. 'The price of these developments becomes evident only now. It is like building a super-highway, a marvel of engineering that permits travelling at 70mph,' he said. 'The only trouble is that this road ends abruptly and no provisions are made to build an exit.'

CHAPTER 2

THE MILLION DOLLAR BOYS' CLUB

The Public Wasn't Ready For 'Burn Baby Burn'

A few days before the *Mobro*'s load was burned, a 41-year-old legislator from Setauket, on the north shore of Long Island, declared war on plastics.

Steven Englebright was spurred on after watching talk show host Johnny Carson poke fun at the *Mobro*'s trash – or, as Englebright saw it, Long Island's trash. Carson had made the barge a feature of *The Tonight Show*, pulling out an enormous map of the world to show his audience where Islip was and where the wandering barge was located.

'Captain Duffy St Pierre, take your barge, hang a left around Florida – hang a hard left – and take off due east until you hit the Canary Islands. Hang another left, and go through Gibraltar. Cross the Mediterranean Sea, down to the Suez Canal, into the Red Sea. Do a U-ey at Yemen. You're now in the Arabian Sea. A hard left at the country of Oman, up into the Gulf of Persia and – there is Iran. Dump it right there. Then next week the ayatollah will probably come out and declare it a state park,' Carson had quipped to wild applause.

Englebright, like many on Long Island, was embarrassed by Carson's hazing. 'It was humorous but I didn't appreciate the irony at all,' he remembers. 'This is my home and I felt responsible.'

The legislator's emotions stretched well beyond embarrassment. He had, for some time now, been worrying about how trash was impacting Suffolk County's fragile water supply. Before becoming a legislator, Englebright – who has a master's in palaeontology – had taught college-level classes about Long Island's geology and water resources at Stony Brook University. He continued teaching part-time once elected, while also serving as director of the Museum of Long Island Natural Sciences.

Although surrounded by salty water, Suffolk County had a reservoir of fresh sub-surface water that had collected over thousands of years in an underground aquifer. Over the decade leading up to the *Mobro*'s voyage, there were unmissable signs that the once-pristine water was being contaminated with chemical and biological waste. Some of this waste was leaching from unlined landfills. Increased reports of miscarriages, breast cancer and neurological issues began cropping up, raising questions about the quality of the water Long Island residents were drinking. 'I had been lecturing not just to my students but to anybody who would listen that I saw a coming crisis with water quality,' says Englebright.

Concerns mounted so much that, in 1983, the state mandated that landfills in the area would have to close by December 1990 to protect the water. That law set in motion expensive trucking of the county's waste to as far as western Pennsylvania, 400 miles away, and garbage burning. Over the summer when the *Mobro* dominated the news, Englebright spent sleepless nights mulling over ideas to lessen the amount Long Island was sending to landfills.

As a young boy, he had spent months on his grandparents' five-acre farm outside of Evansville, Indiana, and been fascinated by a compost pile his grandfather had set up behind the garage. Englebright had helped throw in orange peels, lemons, paper and other organic refuse that would eventually be used to grow vegetables. 'I wondered why we didn't think about using biodegradability and then I realised, *Oh, it's pretty obvious we are using a lot of plastic and it's not biodegradable at all,*' says Englebright. That thought set in motion a chain of events that would impact the plastics industry, not just in Suffolk County but all over the US.

In August 1987, Englebright introduced a bill to ban fast-food restaurants, supermarkets and delis in Suffolk County from using plastic – or any other type of packaging that was not biodegradable. Packaging made up the county's single largest category of waste. Englebright saw banning plastics as a way to cut back on incineration, spur the use of biodegradable packaging and send a strong message to the plastics industry to aggressively begin recycling or face further bans. The proposal, if approved, would create the broadest and strictest packaging regulations of any municipality in the US.

Chemical companies, retailers and restaurant owners banded together to protest vociferously. Ultimately, seven months later, a narrower version of Englebright's bill was waved through by the county legislature. It banned plastic bags and food containers made of polystyrene and PVC – the plastic that released toxic dioxins when burned. Violators would pay $500 each time they were caught breaking the new law, slated to go into effect in July the following year.

Although less far-reaching than his initial proposal, Englebright's new bill still posed a serious threat to the plastics industry and to big customers like McDonald's, a major user of polystyrene foam. Suffolk County had pioneered no-smoking laws and deposits on drinks containers that had caught on elsewhere. Where it went, the nation could easily follow.

*

It was a warm day in Washington, DC in May of 1988 when Donald Shea loaded up his car with the motley collection of possessions he had accumulated as head lobbyist for the Beer Institute. Shea didn't consider himself a sentimental guy, but after nine years fighting on behalf of beer behemoths like Budweiser maker Anheuser-Busch, he allowed himself some small sense of nostalgia, slipping a beer mug and a wooden model delivery truck into one of his boxes.

Shea was headed only five blocks over. His new office stood on the fourth floor of a building on 13th and K Street, overlooking Franklin Park. Listening to Bruce Springsteen pump from the sound system in his snub-nosed red BMW, he tingled slightly with the excitement of this new chapter.

His brief was clear: to turn around the precipitously declining reputation of plastics and forestall a tidal wave of regulation that was already breaking from all corners of the US.

Suffolk County's measures had grabbed the attention of counties and towns across the country and now hundreds of copycat laws were being proposed to curb plastic. Seven chemical giants – Amoco, Dow Chemical, DuPont, Exxon Chemical, Quantum, Cain and Mobil Chemical – were banding together to fight back. Between them, they made an enormous slice of the plastic packaging gleaming on every shelf in every supermarket up and down America.

It was still early days in what promised to be a multi-billion-dollar sales boom driven by America's embrace of throwaway plastic, and the chemicals giants were on the winning end. In 1988, demand for plastic was close to outstripping supply, factories were working full tilt and profit margins were among the fattest they had ever been. Oil prices had been spiralling downwards, making it cheaper to make new plastic. Meanwhile, a weaker dollar was spurring American exports. By now, Procter & Gamble, McDonald's, Heinz, Unilever, Coca-Cola, Pepsi and other consumer goods companies had all turned to single-use bottles, cups, bags, sachets, boxes and pouches to cheaply and conveniently package their products.

The chemical companies had each thrown in a million dollars to fund a new guerrilla group, one charged with systematically tearing down or countering every one of the anti-plastic claims threatening to wipe out billions of dollars in sales.

'They said, "We've got to get our arms around this. We're getting clobbered in the marketplace and customers will start choosing other products",' recalls Shea. 'It was driven, frankly, by the apprehension of plastic bans and severe restrictions burgeoning all over the country.'

At 37, Shea had dark hair, blue eyes and sharply cut features. He was something of a master negotiator. Diplomacy was a trait he had developed partly as a survival instinct as the sixth among eight siblings growing up in Inwood at the northern tip of Manhattan. Shea's parents, both of Irish ancestry, had worked hard to give their kids a good life. His father was an accountant and his mother worked secretarial and administrative jobs. Their two-bedroom, one-bathroom flat was never quiet, filled with the voices of children and the more melodic sounds of Harry Belafonte and Miriam Makeba wafting from a box stereo in the living room. Despite the ongoing chaos of life with four brothers and three sisters, Shea had found the space to lose himself in history books, each morning rising earlier than everyone else to read in his bunk bed.

After graduating from college with a degree in English literature, Shea taught English and history at a school in the Bronx for a couple of years. He then dabbled in various sales and marketing jobs before deciding to go to law school at night. He graduated with honours but chose not to practise. By then, he was already working for the beer industry and found the scrappy, itinerant life of a lobbyist far more thrilling than the confines of any law firm.

The Beer Institute had been around since 1862, when it was set up to fight federal taxes levied during the American Civil War. Shea had successfully deployed tactics drawn from a much-used playbook to crush a proposed ban on beer ads, stop higher taxes and quell public outrage about rising drunk-driving cases.

All this had earned him a reputation on Capitol Hill as one of the better lobbyists around. But a restructuring three years earlier at the Beer Institute had diverted much of the industry's lobbying in-house to individual companies, turning Shea into a consultant for the revamped trade body with too much time on his hands. As he hurtled towards middle age, fighting the same sud-soaked battles that beer lobbyists had grappled with for decades felt less thrilling than it had once been.

When a head-hunter for the yet-to-be-formed plastics lobby group approached him, Shea didn't need much explaining to grasp the reputational challenge the chemical companies were facing. At home in DC with his wife and two daughters, Shea had watched riveted – like so many other Americans – as the nightly news showed footage of the *Mobro* being turned away over and over again. 'You literally had no choice,' he remembers. 'If you watched the evening news, it was on.'

What the chemical giants were offering – apart from a lot more money – was the chance for Shea to once again be in the driver's seat. 'We would make our own history,' he says.

By the mid-1980s, the US was using more plastic than steel, copper and aluminium combined, and the plastics industry's annual revenue stood at $138 billion. Littered plastic was showing up across beaches, sidewalks, roads and parks. For the first time, many Americans woke up to the idea that a fast-food box or a plastic straw used for a few minutes could hang around for hundreds of years. 'America was running out of landfills and plastics was the product being blamed,' says Shea. 'I think, frankly, that's because of its ubiquity. You were reminded of plastics everywhere you went.'

In Shea's first year on the job, Englebright's bill provoked around 500 different pieces of state and municipal legislation looking to restrict or ban plastic. 'Suffolk County started and then all of sudden every person who ever put a bumper sticker on their car was anti-plastic,' remembers Don Olsen, who worked as head of corporate communications for Huntsman Chemical, a big polystyrene supplier. 'It really caught the industry by surprise when all this backlash came our way.'

Close to a decade defending brewers had left Shea with a few tricks up his sleeve. The liquor industry had long fought the perception that a shot of vodka or whisky was more intoxicating than a bottle of beer or a glass of wine. Much to the beer industry's chagrin, big liquor's campaign to convince legislators that a standard shot didn't contain any more alcohol than a standard bottle of beer – and shouldn't be taxed any higher – had found many sympathetic listeners.

Shea took a leaf out of the spirits industry's book. He set out to diminish plastic's perceived contribution to America's waste problems by showing how other types of packaging were just as bad, if not worse for the environment. 'My task was mainly a political and PR challenge,' he says.

To prove his point, Shea commissioned William Rathje, a Harvard-educated anthropologist, to show that if plastics were clogging up landfills, they weren't alone: so were paper, food and purportedly 'degradable plastics'. The latter were being pushed by grain processing giant Archer-Daniels-Midland Company. It saw the potential for a new billion-dollar market for corn and was loudly telling anyone who would listen that adding corn starch to traditional plastic could make it degrade.

Heavy-set with a closely trimmed salt-and-pepper beard, Rathje wore sunglasses that dangled from a string around his neck and called himself a 'garbologist'. He spent his days digging in landfills and analysing the contents. Once consigned to obscure academic journals, Rathje's research was now in high demand from companies like

Pampers maker Procter & Gamble as regulators considered banning non-degradable diapers. Rathje was as unafraid of broadcasting his fascination with trash as he was of pulling on a pair of black gloves to dig through piles of foetid leftovers. 'Bill was a delightful guy, but I never liked to be around him,' says Shea. 'He stunk.'

Rathje's finds from excavating landfills included legible newspapers, a ten-year-old cup that said 'This cup is biodegradable' and 20-year-old corn cobs with the kernels still on them. He pointed out that if bugs weren't eating the corn, they were very unlikely to eat the corn starch used to make 'degradable' plastics. Shea's lobbyists used Rathje's findings as evidence before lawmakers that nothing degraded in landfills and plastics were being unfairly singled out.

Being overtly seen as pursuing recycling was another prong of Shea's attack plan.

While at the Beer Institute, he had gone to great lengths to show that brewers were voluntarily spending large sums of money to discourage drunk driving. He funded ads with punchy messages like 'You drink, you drive, you die' and was a founding board member of Students Against Driving Drunk. 'I was one of the people instrumental in preventing beer advertising from being banned,' he says with more than a hint of pride. 'And we didn't get an excise tax increase.'

If the plastics makers wanted to forestall bans and taxes, they needed to be very publicly seen to embrace recycling, says Shea. At the time, only 1 per cent of plastics were collected for recycling, compared to 29 per cent of aluminium, 21 per cent of paper and 7 per cent of glass.

Many plastics makers still favoured burning plastic and capturing any energy that could be produced in the process. Shea knew that incineration, from a public relations standpoint, was a no-go zone, given the studies that had shown chlorine-containing plastics could release deadly dioxins when burned. 'Part of our mindset in 1988 was that the public wasn't ready to hear the plastics industry say "burn baby burn",' says Shea, 'but we always thought it was going to be a key component.'

The new plastics lobby group headed by Shea was stuffed with the most profitable chemical makers in America. It was called the Council for Solid Waste Solutions, a name carefully chosen to avoid drawing attention to plastics. CSWS's logo consisted of a globe cradled lovingly between a pair of hands. Given its members had each thrown in a cool million, the industry nicknamed it 'the Million Dollar Boys' Club'.

The Council had its own board made up of top executives from each of the seven companies and held monthly wargaming sessions in hotels dotted around Washington, DC. At one of these, Shea and his newly hired three-person team unfurled a life-sized poster showing a car attempting to make its way down a road replete with different roadblocks: a stop sign, a speed bump, a pothole. The car stood for CSWS, while the hurdles represented things like negative public opinion and recycling. The point, says Shea, was to convince the chemical company executives that focusing just on incineration was akin to driving down a road to nowhere.

'The board agreed that, no matter what our message is, if we don't make efforts to develop plastics recycling, we won't have the bona fides necessary for people to listen to us,' says Shea. 'We were set up to make sure that bad bills didn't get passed. And we were set up with the understanding that for that to occur, the industry would communicate or develop what were perceived as solutions.'

*

After breakfast one morning in early July 1988, Ronald Liesemer backed his maroon Oldsmobile Cutlass Ciera out of his driveway in Wilmington and headed across the Delaware Memorial Bridge. He drove 90 miles to Rutgers University in New Jersey. Upon arrival, he asked around until he found his way to the Center for Plastics Recycling Research.

Liesemer was DuPont's head of product development for plastics and elastomers. In contrast to the blustering personalities who made up the top ranks at the chemical companies, the 50-year-old was quiet, earnest and polite. He wore well-pressed suits and rarely left the house in shoes that weren't polished to a shine.

Liesemer had recently been seconded to CSWS and his new job was to show the public that plastics could be recycled. Many counties and cities across the US were designing their first recycling programmes, ordering trash cans and trucks with compartments for different materials. Plastics that weren't included in these programmes shot to the top of many states' ban lists.

By 1987, roughly a third of DuPont's $1 billion research budget was being spent on plastics and the company was turning close to $4 billion in sales from the material. But all that could vanish if plastics didn't start being recycled. 'There was legislation to ban any plastic that was not recycled,' explains Liesemer. 'So the answer to that was recycling. If we didn't get in quickly, we would be designed out.'

Back in 1985, the Society of the Plastics Industry – a long-time trade group which counted more than 2,000 members – had banded together with container makers and soda giants Coca-Cola and Pepsi to fund the recycling research hub at Rutgers University. Coke at the time was trialling a clear 12-ounce plastic can with an aluminium top, a first for the soft drink industry. Consumers liked the plastic can. But ultimately – after opposition from environmentalists who said burning the can would release dangerous fumes – Coke decided not to roll it out further unless it could be recycled. Recycling was becoming a get-out-of-jail-free card in a situation otherwise riddled with reputational risk.

Three million dollars was earmarked over three years for the Center for Plastics Recycling Research's efforts to turn used plastics into lumber boards, pillow stuffing and floor tiles. At first, the Rutgers researchers focused on the two most recognised – and hence easily sortable – plastic containers: polyethylene terephthalate (PET) soda bottles and high-density polyethylene (HDPE) milk jugs. They devised a system to grind and separate the two plastics, strip their labels and filter out the aluminium caps. They sold plastics recycling machines to entrepreneurs for just $3,000.

But recycling soda bottles and milk jugs didn't help with the dozens of other types of plastic, increasingly being used in combination with each other. In a prescient observation, the marketing head for GE Plastics – a division of General Electric – warned in 1987 that referring to the panoply of different materials simply as 'plastic' would only hinder recycling efforts. 'We have to be more critical of ourselves in terms of what plastics should be used and how they should be recycled,' said Uwe Wascher. 'The industry must stop thinking of itself as a one-product business.'

The Rutgers researchers' next big project was trying to recycle mixed plastics into items such as fence posts and parking bay barriers. 'Rutgers was the only game in town,' says Liesemer, who had high hopes for what he'd find there. But when he arrived, he saw rows of large blue recycling boxes, each marked for paper, glass, aluminium and plastic. 'I saw that and thought *This is not the way the world is going to run. Households – even if they put these containers in the garage – won't tolerate it, and we'd never get them in apartment buildings.*'[1]

---

[1] Wascher and Liesemer had both hit on issues that plague the plastics industry to this day. Although people talk about plastics like a single material, it isn't one. For the most part, this versatility makes it uneconomical to recycle. To deal with people's reluctance to house bins for various items, many municipalities ultimately allowed residents to throw all recyclables in one bin. Such 'single-stream' recycling saves space, is more convenient and can make collection cheaper. But it's more expensive to sort and more susceptible to contaminants sneaking in, destabilising the already fragile economics underpinning recycling.

Liesemer also inspected the lumber made from recycled mixed plastic, immediately recognising it as very poor quality. Different plastics melt at different temperatures, and the researchers' inability to sort plastic by type left some plastic pieces whole in the lumber. Others had been heated too much, decomposing and leaving black spots. 'What they had would never be saleable,' recalls Liesemer. 'It was ugly and warped.'

Even researchers inside Rutgers could see the shortcomings of their recycled plastic. 'In the words of Henry Ford, you could get any colour you wanted as long as it was black,' quipped one assistant professor who worked at the research centre. Despite this, the centre's director publicly remained optimistic, predicting that plastics would 'go from being banned and bashed to being the most recyclable and recycled material in the waste stream'.

In time, the technical issues hindering plastic lumber would be sorted out but, even today, mixed plastics recycling is plagued by collection and sorting problems; the resulting park benches and picnic tables have never commanded a high-enough price to justify collecting and cleaning big volumes of often highly contaminated plastics.

Back in the 1980s, despite the chemical companies publicly touting recycling as the answer to plastic waste, many knew that, for most plastics, recycling was unlikely to amount to much more than a public relations exercise. 'We honestly recognised early on that this was a terrifically complex issue influenced by lots of factors, not the least of which was the plethora of types of plastics,' says Dennis McKeever, who as Dow's vice president for polyolefins and elastomers was also its representative on the Council for Solid Waste Solutions.

McKeever, who had a PhD in physical chemistry, belonged to the camp who thought incinerating plastics and capturing some energy in the process was the best solution. 'The economics reared its head quickly when we looked at how we collect and separate, and which of these plastics would be pragmatic to economically recycle,' McKeever remembers.

Liesemer himself had serious doubts that plastics recycling could work. 'Within the industry, I don't think there was ever a sense that a high percentage of plastics would be recycled,' he says. He recalls telling a CSWS board member that hitting a recycling rate of 10 per cent for all plastics would be a big achievement, to which the man replied irritably, 'Well, why are we doing any of this then?' 'I told him,' says Liesemer, '"because we have to *prove* to the public that it can be done".'

In other words, recycling was a giant marketing campaign.

*

In 1989, DuPont announced it would partner with the nation's largest waste hauler to open a network of five recycling plants, while another plastics maker, Wellman, set up a joint venture with waste disposal giant Browning-Ferris Industries. Dow struck a deal with Domtar of Canada to recycle plastic soda bottles and milk jugs, and Amoco opened a plant in Brooklyn to take in foam fast-food containers, pledging to use the pellets to make wall boards for construction. CSWS deployed staff to convince municipal governments to add plastic soda bottles and milk jugs to their kerbside recycling programmes. More recycling programmes for plastic were established in 1990 alone than had been in the previous ten years combined.

The flurry of recycling attempts was evidence of the deep anxiety rippling through the chemical giants as public perceptions about plastics continued to deteriorate. A memo to the industry from the president of the Society of the Plastics Industry in December 1989 warned that the image of plastic has 'plummeted so far and so fast, in fact, that we are approaching a "point of no return"'.

By 1989, 72 per cent of Americans polled said they believed plastics were harmful to the environment, up from 56 per cent just one year earlier. 'At this rate we will soon reach a point from which it will be impossible to recover our credibility,' wrote SPI president Larry Thomas. 'Business is being lost. Product growth rates are being dampened. And stock analysts are beginning to take notice.'

CHAPTER 3

McTOXIC

'What you may not know is that polystyrene can be recycled'

Of all the products that came to be seen as symbols of America's profligacy following the *Mobro*'s infamous voyage, McDonald's bulky foam clamshell container was the most prominent.

McDonald's had relied on disposable packaging since 1948 when Richard and Maurice McDonald fired all the waitresses at their McDonald Brothers Burger Bar Drive-In restaurant in San Bernardino, California. They rolled out an entirely new model of serving food designed to be high-volume, very cheap and very fast.

Central to that new way of working was replacing the silverware and china plates with disposable bags, cups and plates made from paper. By getting consumers to just throw cups and other packaging away, the McDonald brothers had neatly rid themselves of the cost and hassle of washing and drying hundreds of pieces of dishware each day. 'Imagine – No Carhops – No Waitresses – No Dishwashers – No Bus Boys – the McDonald's System is Self-Service!' went one early ad aimed at potential franchisees. Labour costs plummeted, sales soared and soon the brothers were pulling in $100,000 a year in profits from the restaurant, a tidy sum in the 1950s.

When Ray Kroc – who sold paper cups and milkshake mixers – bought the rights to franchise McDonald's nationwide, he naturally followed the same model. McDonald's became one of the largest users of paper packaging in the country.

But the paper containers didn't retain heat well, which made them less than ideal. Through the 1960s, worries about deforestation were also mounting. Studies showed that paper made up as much as 60 per cent of highway litter and half of what people threw away in big cities. In large part to allay environmental concerns, McDonald's, ironically, decided to switch away from paper to plastic containers for its Big Macs. On 22nd September, 1975, McDonald's rolled out the polystyrene clamshell container across the US. The boxy foam container replaced the foil and paper used to wrap its burgers, which had also been secured by a paper collar and placed in a red paperboard box.

The fast-food giant's success boiled down to a simple formula. It speedily served consistently tasty, hot food at low prices. The polystyrene clamshell, developed by Salt Lake City-based Huntsman Chemical, was the perfect match. Staff could quickly layer sandwiches inside the box and easily close it, the food stayed warm and tasted fresh, and – perhaps most importantly – at less than 2 cents, the container was cheap.

'When the company went to Styrofoam, everyone loved it,' says Shelby Yastrow, who started doing legal work for McDonald's in 1960 and later became its general counsel and head of environmental affairs.

As if the clamshell wasn't good enough, in 1985 McDonald's launched a double clamshell to house a new sandwich called the McDLT. The McDLT was essentially a burger with lettuce and tomato. When McDonald's product development department created it, taste tests had shown there was a problem. The cold lettuce and tomato quickly cooled the burger down, detracting from the sandwich's flavour.

'One thing that I learned over the years that's beyond dispute is that hot is important,' Yastrow, now retired and living in Arizona, says. 'When you have taste tests and you give someone 12 burgers, they always pick the hottest one.'

The double clamshell neatly solved this problem. The new package had two compartments: one for the hot components – the burger and the bottom half of the bun – and the other for the lettuce, tomato, cheese, sauces and the top of the bun.

The company's hard-charging US president Ed Rensi seized on the new sandwich in its fancy new container as McDonald's chance to wipe the floor with competitors like Burger King and Wendy's. It's estimated that he poured $100 million into advertising the McDLT in the first few months of its launch.

One 1985 TV ad showed a dancing Jason Alexander – who went on to star in *Seinfeld* – singing 'The beef stays hot, the cool stays crisp / Put it together you can't resist.' Another featured a pristine white double clamshell with one compartment resembling an oven and the other a refrigerator. 'Its two-sided package keeps the burger hot on the McOven side and the McFridge side keeps the lettuce and tomato cool and crisp,' a woman gushed in the soothing tones of a massage therapist.

By 1987, McDonald's had cornered 36 per cent of the US fast-food market, up two percentage points from two years earlier. That might not sound like much, but it is when the market is worth $25 billion.

To give you a sense of its scale: McDonald's alone that year bought 120 million pounds of lettuce and sold nearly one out of every three hamburgers in the US. The company's logo adorned

7,500 US restaurants and thousands more overseas. Globally, a new McDonald's restaurant was opening every 17 hours.

*

Made from the chemical styrene, polystyrene was, by the 1960s, the third most widely used plastic in the US. It was fast elbowing out paper for meat trays, egg cartons, cups, plates and cottage cheese tubs. By the 1980s, around 3,000 tons of polystyrene were ending up in North American trash cans every day, enough to make 900 million coffee cups or 100 75-storey office towers, reported the *Globe and Mail*. And by the time the *Mobro* set sail, McDonald's was the largest user of polystyrene foam in the US.

McDonald's was using polystyrene foam not just for burgers but for breakfast items, drinks and salads. Often the distinctive foam packages ended up as litter. In the months after the barge's contents were burned, Richard Kessel, the head of New York state's consumer watchdog, teamed up with the city's mayor Ed Koch to publicly press McDonald's to stop using the containers. 'We are going to be buried in our garbage,' Kessel warned. 'We're going to be wrapped up in packaging.'

New York City had been one of the earliest places to single out plastic, in 1971 implementing a tax on plastic containers. The tax didn't last because a judge ruled it discriminatory, but it was the start of the city's fractious relationship with plastic litter.

At a November 1987 press conference on Long Island, Kessel waved around an empty McDonald's burger container like it was the worst thing he'd ever seen. Styrofoam,[1] he said, was not biodegradable, produced harmful emissions when burned and took

---

[1] Styrofoam is a brand name for polystyrene foam.

up more space in landfills than paper alternatives. He was asking McDonald's, as the biggest user of Styrofoam containers, to set an example by switching to recyclable paper. Should his request be ignored, Kessel indicated that a ban could be on the horizon.

Over in Suffolk County, Steven Englebright had just proposed his anti-plastics bill. In setting out to ban polystyrene fast-food packaging, he had squarely taken aim at McDonald's. 'I just thought this was something that had to be addressed,' he recalls. 'Publicly owned landfills were paid for with taxpayer dollars and so the costs of disposal for plastic waste were not being borne by the industry but instead passed on to citizens.'

In attempting to lower costs for taxpayers, Englebright was raising a key issue, one you will see repeatedly highlighted throughout this book. Why were taxpayers picking up the expense of sending bulky plastic packaging to landfills even though McDonald's was the one choosing to use it?

Ed Rensi had no intention of abandoning the clamshell. Loud, brash and impatient, the McDonald's president had a fiercely competitive streak and bristled at what he saw as environmentalists and politicians telling him how to run the company he had worked at for his entire adult life.

Rensi first applied for a job at McDonald's in 1966 after seeing an ad in the window of a local restaurant across the street from Ohio State University, where he was studying education. Back then, he had never heard of the chain, despite the 600 restaurants that carried its name. He took the job to make money to support his young family; his wife Anne had just had a baby girl.

Having grown up in a family that owned a restaurant, quickly grilling burgers came naturally to Rensi. Colleagues soon nicknamed him 'Fast Eddie'. He worked his way up to trainee manager, eventually landing a job as vice president of operations based in the fast-food giant's Oak Brook headquarters in suburban

Chicago. He became president and chief operating officer of McDonald's US in 1984.

Rensi was a workaholic, spending three to four days a week up to 30 weeks a year visiting McDonald's restaurants across the country. He checked that the burgers were hot, the tables were wiped and that customers were well taken care of. He didn't brook laziness. 'If you've got time to lean, you've got time to clean,' he'd say.

He was fiercely loyal to McDonald's, eating breakfast and lunch there most days in an apparent attempt to thumb his nose at critics who said fast food raises cholesterol. He's still around and very active all these years later, currently serving as co-chairman of the Boardroom Initiative, a group that describes its mission as defending free-market capitalism, including pushing people to divest from corporations that pursue 'extreme environmentalism' and participate in 'woke cancel culture' or 'unhinged social justice' over running a 'good business'.

Once Rensi became president, his go-to man on major reputational matters, as the clamshell was fast threatening to become, was 51-year-old Shelby Yastrow. Yastrow's relationship with McDonald's had begun back in 1959. Just two weeks out of Northwestern's law school, he was ambling past a conference room at the large Chicago law firm he had newly joined when a senior partner imperiously beckoned to him. 'Young man, come in here,' he called.

The partner introduced Yastrow to Ray Kroc. Kroc was fighting an Illinois franchisee group that had opened a successful McDonald's in Urbana and wanted to open another in Peoria. When Kroc said they couldn't because he had already promised the location to someone else, the men opened a McDonald's lookalike, called Sandy's. 'It was a carbon copy,' says Yastrow.

Kroc took the men to court, accusing them of violating a non-compete clause in the contract they had signed and of stealing McDonald's trade secrets. Yastrow got assigned to the Illinois case, even though he hadn't yet passed the bar. As the youngest employee at Sonnenschein Carlin Nath & Rosenthal, he initially just answered phones and wrote memos. 'I didn't know anything. I was a jerk,' says Yastrow. 'I was kind of just a gofer, a little punk.'

As the case dragged on, the senior partner in charge left Yastrow to make the trips to Peoria and shoulder much of the load. Much was at stake. 'If Ray lost, it meant any franchisees could go after him and build their own restaurants,' says Yastrow. 'It was a big deal to him and a big deal to me.' The case went all the way up to the Illinois supreme court. The parties eventually settled, agreeing that the Urbana McDonald's would close but that Sandy's could remain.

In the years that followed, Kroc and Yastrow struck up a friendship. They spent many hours together during trips on Kroc's small private plane to the towns and cities in which new McDonald's restaurants were grappling with zoning or signage disputes. Eventually, in 1978, Yastrow, by now 42 years old, started working at the fast-food chain full time, walking away from the small Waukegan, Illinois law firm he had left his Chicago employer to open. At McDonald's, he led a team that would grow to more than 100 attorneys and fight cases in courtrooms not just across the US but also in Brussels, Frankfurt, Tel Aviv and Sydney.

Yastrow had dark hair he slicked into a neat side parting, wore his reading glasses on the end of his nose and favoured pin-striped suits. He had a ready laugh, used expansive gestures and loved spinning yarns – so much so that he wrote novels in his spare time, which went on to be pretty successful. Everyone called him 'Uncle Shelby'. The nickname came about after he fell into the habit of getting past the formidable secretaries of high-flying lawyers by telling them that 'Uncle Shelby' was on the line. He was popular among his colleagues and revelled in being seen as the polar opposite of Rensi.

Yastrow got roped into defending the clamshell after Rensi one day cornered him in the men's room. The McDonald's president knew that Yastrow had met New York governor Mario Cuomo a few times. Now the state was considering following Englebright's lead and banning all polystyrene foam containers. Rensi wanted Yastrow to get Cuomo on-side and make the whole overblown mess go away. The McDonald's president's orders to Yastrow were unequivocal: fight to keep polystyrene, whatever it takes.

'The only thing I knew about landfills in those days was when I looked at my son's bedroom,' says Yastrow, who told Rensi as much. 'But I took it seriously and I became interested in green issues.'

*

In October 1987, protesters all over the US flooded into their local McDonald's, arriving at peak lunch hour with their sons, daughters, nephews and nieces. 'I'll have a Big Mac but I'll take it on a paper napkin, no Styrofoam please,' was the order politely given by each person who stepped up to the counter. Some asked for nuggets or other menu items, but they all wanted their orders on paper napkins.

The grassroots effort had been spearheaded by the Citizens' Clearinghouse for Hazardous Waste, a non-profit organisation that had sprung up in the wake of the Love Canal disaster in which families unknowingly living on a former chemical dump suffered from cancer, miscarriages and other serious health issues. CCHW was focused on connecting communities that were fighting dumps, incinerators and hazardous waste problems so they could pile pressure on governments and businesses.

CCHW's organising director, Will Collette, idolised the Chicago-born social movement organiser Saul Alinsky, who openly courted controversy as a way to effect change. 'Alinsky introduced me to the idea that you can be a professional trouble-maker,' says Collette. 'Being a hell raiser is a lot of fun.'

A chain-smoker with reddish brown balding hair and large glasses, Collette often flew around the country to galvanise communities who didn't want new landfills or incinerators sited near their homes. The hearings and rallies he spoke at could get pretty hairy – Ku Klux Klansmen sometimes appeared; on several occasions, Collette was forced to leave town with a police escort. He usually booby-trapped his hotel room door to detect break-ins.

CCHW was perpetually short on funds, operating out of the cockroach-infested second floor of a shopfront in Arlington, Virginia. Staff called it 'the roach motel' but tolerated the scurrying inhabitants; the dingy surroundings reassured visiting donors that the non-profit was serious about its work and not trying to line its own pockets.

Collette saw taking on a behemoth like McDonald's, which by then was pulling in close to $5 billion in revenue a year, as the best way to overcome CCHW's limitations. Its staffers harnessed a strategy called mass jiu-jitsu: using an opponent's weight against them. 'We had a very sharp focus on who's got the power and who's doing this to you,' says Collette. 'Your mission is to become such an irritant in their life that they will reach the point where they'll stop what they're doing.'

To maximise irritation, Collette and the other activists had galvanised their networks of supporters, which included lots of young mothers, for a 'national action day' against McDonald's. Asking for burgers without any plastic was aimed at causing maximum disruption. 'It was the kids who were the shock troops in the campaign,' says Collette. 'They weren't doing anything illegal so restaurants couldn't call the police.'

In the months before the national action day, CCHW and its network of fellow non-profits had begun sending letters to McDonald's, warning that if the company didn't stop using polystyrene it could expect a highly visible nationwide grassroots fight that would centre around its restaurants. Yastrow took the letters seriously. He flew from Chicago to Washington, DC to meet Collette, who insisted their meeting take place in an open hotel lobby. It was a tactic designed to discomfit the general counsel and his team of lawyers. The two men and their accompanying staff – about a dozen people in total – huddled in their groups on two sides of a round wooden table. Collette unfurled an enormous poster showing the Statue of Liberty engulfed by clamshell containers, artwork for a proposed campaign.

He began listing the reasons McDonald's should abandon polystyrene. It was produced using ozone-depleting chemicals. It was made from styrene; the chemical had been linked to cancer and had turned up in the fat tissues of factory workers and even consumers, likely transmitted by hamburger fat and drinks. Burning polystyrene could also release cancer-causing chemicals,[2] said Collette. And if all that wasn't enough, McDonald's clamshells contributed 1.3 billion cubic feet of foam waste each year in the US alone, he added.

'Don't make us do this,' he warned Yastrow.

At its heart, McDonald's had always tried to follow consumers. It added chicken nuggets when Americans began eating less beef. It introduced salads and cut the sodium content of its pickles and cheese as people worried more about salt. It converted from animal fat to vegetable oil in an attempt to mitigate concerns about heart health. It was clear that consumer sentiment against the clamshell was turning, and yet it wasn't obvious what the way forward should be. Yastrow wanted to find a way to keep the clamshell but also please its critics.

Polystyrene was made with chlorofluorocarbons, chemicals that damage the ozone layer which protects us from the sun's cancer-causing ultraviolet rays. The CFCs were increasingly linked to a warming climate. By now there was broad consensus that there was

---

[2] In the course of my reporting, I didn't find studies showing this. Researchers' concern about toxic fumes from controlled burning of plastics focused mainly on polyvinyl chloride.

no way to avoid a significant rise in temperature in the coming century. A few scientists had begun linking throwaway living with climate change. For the most part, though, plastics in the late 1980s weren't widely linked with climate change the way they would be in the years ahead, when the public would focus more on plastics' fossil-fuel origins.

Yastrow's first bid to appease the environmentalists was to announce that McDonald's would stop using CFCs and switch to a different chemical.[3] Even as the chain made the concession, it was careful to downplay its own role. It said it didn't think the move would have any real impact on the level of CFCs being emitted, given how tiny its contribution was, and it was only making the change to set an example others could follow.

Pledging it would eliminate the CFCs had done little to assuage CCHW's main concerns. So, at the Washington, DC meeting, Yastrow offered another solution – recycling. While the technology to recycle polystyrene foam already existed, it was used only for factory-floor scraps. McDonald's, Yastrow said, was exploring ways to recycle all the clamshells used in its restaurants.

But Collette's concerns regarding McDonald's revolved not so much around reducing waste but around the production process for polystyrene, which he believed was toxic. He wasn't ready to settle for 'namby-pamby efforts towards recycling' or anything less than a full abandonment of the clamshell. 'We weren't expecting anything and we didn't want anything,' he says. 'We were ready to fight.'

---

[3] McDonald's was accused of still using a CFC – just a different sort. After lobbying from manufacturers, the Environmental Protection Agency changed CFC-22's designation to HCFC-22, allowing McDonald's to make the CFC-free claim. Environmentalists argued that the new agent still depleted the ozone layer, although less so. The allegations are made in a 1989 opinion piece: 'McTruth: Fast Food for Thought', Curtis A. Moore, *Washington Post*, 10th December, 1989.

Back in Arlington, CCHW staffers created a 'McFact pack' listing the drawbacks of polystyrene. Polystyrene, it said, generated the fifth largest amount of toxic waste of any single chemical production process. CCHW mailed its claims widely – to journalists, environmental organisations, churches and community leaders.

The allegation that McDonald's was destroying nature started to get picked up by the press. 'It was a nightmare for us,' says Yastrow. 'I would go to cocktail parties or friends' homes for dinner and that's the first thing that would always come up.'

The national action day, which came two months after Yastrow's meeting with Collette, caused all the disruption Collette had hoped. After a day of fielding calls from overwhelmed franchisees and frazzled restaurant managers, Yastrow shifted gears. He used a tactic employed by corporations and their lobbyists the world over – and one you will see repeatedly used by companies throughout this book: minimising the problem. McDonald's public relations team was briefed to tell anyone who called that polystyrene made up just a small fraction of America's waste.

Yastrow also hired a Kansas-based consulting firm, Franklin Associates – which regularly worked for the plastics industry – to show that polystyrene was far less taxing on the environment than paper because it was more than 90 per cent air, making it less energy-intensive to produce.

By then, word had also spread about the work that Bill Rathje, the garbologist, was doing for the Council for Solid Waste Solutions. McDonald's began flying Rathje to legislative hearings so he could explain how paper was clogging up landfills. Rathje, through his digs, estimated that polystyrene foam made up just 1 per cent of what was going into landfills.

McDonald's also argued, head-scratchingly, that having its clamshells – four billion of which were discarded annually – sitting in landfills was a good thing, since it helped 'aerate the soil'.

The double clamshell had two compartments: one for hot components and one for cold ones. Image courtesy of Bobby Beauchesne

For a while, Yastrow was convinced that the best way to manage clamshell waste was incineration, and McDonald's set about testing incinerators at some of its restaurants. When I first approached him for this book, Yastrow insisted that McDonald's never tested incinerators. 'I can assure you that your information about incineration is totally bogus,' he wrote. 'McDonald's NEVER "tried incineration for a few months" as you claim. McDonald's never advocated or engaged in the incineration of trash.' After I presented him with a permit for an incinerator given to McDonald's in 1988 by the state of Illinois, and some old news articles about the incinerators, Yastrow – who is now in his late 80s and can certainly be forgiven for not remembering everything that happened more than three decades ago – backtracked. 'One or two franchisees tried incineration on their own,' he said.

Back in 1989, Yastrow was quoted in news outlets as being directly involved with McDonald's incineration efforts. In April of that year, for instance, he told *Restaurant Business*, a trade magazine, that McDonald's was testing incinerators that burn garbage at 1,500 degrees Fahrenheit at two restaurants in Oklahoma and Illinois, and taking the remaining ash to a dump nearby. The smoke met the air pollution guidelines for all states except New York and the ash and smoke were both 'pretty innocuous', Yastrow told the publication. 'An incinerator could technically handle all of a unit's trash, so I would think that this is a good way to go,' Yastrow was quoted as saying. He told *Restaurant Business* that McDonald's was examining the possibility of building a larger incinerator that could take the trash from 400 restaurants.

When we speak over the phone following our email exchanges, Yastrow says he still doesn't remember McDonald's being directly involved in incineration, but he does remember thinking at the time that the method would be a good solution. 'The only downside was that some of the residue had heavy metals in very minute traces, like cadmium,' he tells me.

The incinerator tests were short-lived. Environmentalists got wind of them and created an uproar. Yastrow arrived at the same conclusion that Don Shea had: public opinion against incineration was far too strong. 'We were already fighting with one arm behind our backs,' he remembers. 'We wanted to show everyone we were clean and green and incineration would have been counterintuitive.'

McDonald's saw itself as part of the fabric of the nation's culture, a dependable, wholesome brand grounded in family values, as American as its own apple pie. One survey showed that 96 per cent of American schoolchildren could identify Ronald McDonald, ranking the chain's mascot just behind Santa Claus. About a third of its 20 million weekly customers were children.

Being accused of destroying the environment didn't sit well with that image. Gallingly, Burger King, which packed its burgers in paper, had begun using the situation to its advantage, touting its concern for the environment.

Yet swapping the polystyrene container for paper packaging, which McDonald's had consistently been bad-mouthing, would be a huge climbdown. Switching to paper from polystyrene would also potentially cost double at a time when McDonald's was already grappling with higher costs from rising beef prices and minimum wages. The Franklin report had also given McDonald's a new environmental justification for sticking with plastic.

Yastrow laid out McDonald's stance on paper in another interview with *Restaurant Business* in May 1989. 'We can put the McDLT in paper, but it won't stay as hot, it won't stay juicy, it won't be as convenient or portable and the paper weighs just as much in the landfill as plastic.'

*

In early 1989, a 16-year-old boy named Kurtiz Schneid began dressing up in a clown costume to protest against McDonald's continued use of polystyrene foam.

Schneid hailed from the New Jersey town of West Milford. He and his high school classmates had sat in on a social studies lesson in which their teacher told the class how the Americans once boycotted wool from England to protest against increased taxation. The teacher encouraged the class to consider their own boycott as a way to effect change in an area they cared about. One of Schneid's classmates suggested advocating for an end to Styrofoam. The next thing you know, the class had zeroed in on polystyrene foam trays, which were used in the school cafeteria, lobbying the school's board to replace foam trays and cups with paper ones. The board demurred, saying students wouldn't want to pay the extra five cents for paper containers.

To prove them wrong, Schneid organised a three-day test in which the cafeteria offered lunch on plastic for $1.20 or on paper for $1.25. Most students paid the extra 5 cents. 'My friends and I were handing out nickels,' Schneid confesses. 'I was a big guy for my age and had jobs lawn mowing and leaf raking so I had my own money.'

The school board capitulated. Its abandonment of foam was widely covered in the local news. New Jersey's top environmental official came to congratulate the students in person. Over the next few weeks, Schneid and a few other students wrote letters to 700 schools across New Jersey, asking them to nix foam. They began looking for targets beyond schools too.

Schneid had never thought of himself as an environmentalist, but growing up he had hated the litter strewn around a swimming hole in the woods near his house. Local teenagers left beer cans and other garbage behind and, after a swim, Schneid would return with trash bags and clean up.

He initially set his sights on convincing the small delis and bode-gas in West Milford to stop putting their coffee and sandwiches in plastic foam containers. But the outlets were all owned by different people and it was hard to get dozens of small business owners to change. And so Schneid began pressing McDonald's to abandon foam. He helped organise protests and rallies at which students brandished signs saying 'Styrofoam Must Go'.

By now, CCHW had sent sticky labels to its mailing list of about 20,000 people with the home address for Ray Kroc's widow, Joan, and the corporate address for Shelby Yastrow printed on them. Recipients were asked to mail back their used containers since McDonald's was claiming it wanted clamshells to recycle. 'I heard third-hand that Joan was not happy,' says Will Collette gleefully.

The West Milford students got wind of the non-profit's efforts and began their own send-it-back campaign. Teenagers dug through public trash cans and picked up litter, amassing hundreds of old McDonald's foam containers. Schneid dressed as a character he named 'Ronald McToxic'. He wore a friend's clown costume left over from Halloween, with a blue wig. He sprawled in a shopping cart, his huge red clown feet dangling over the sides, buried under piles of used containers to be 'returned' to the local McDonald's owner.

The students also wrote hundreds of letters. 'When you get school kids writing letters saying, "Dear Ronald McDonald, please stop using plastic", that sends a message,' says Yastrow. 'We were desperate, we were doing everything we could and this was a high priority for the company mostly because of public perception. We liked to be liked and respected.'

The local press, ordinarily forced to write about legislative hearings and overspent budgets, lapped up the student-led action. 'You'll get the chicken. Your great-great-great-great-grandchildren will get the wrapper,' wrote a journalist at *The Record*. 'Think about that the next time you buy McNuggets.'

The West Milford students also began handing out surveys at shopping centres asking people whether they preferred polystyrene, paper or reusable containers. CCHW said it would publish the results in its newsletter. The survey – advertised as being conducted to devise a 'plastics pollution solution' – was undoubtedly leading, but Yastrow's response was telling. 'I know about surveys,' he told a reporter at the time. 'I know how easy it is to get the answers you want.' He continued: 'It's a joke, it's garbage. Give us this much credit: We're not going to put our food in packaging if we don't think our customers want it.'

Ultimately, of the 2,000 people the students polled, 1,839 said they preferred paper hamburger wrappers, while just 82 people favoured polystyrene. Notably, nobody gave reusables much thought.

As the students organised more protests outside restaurants, McDonald's fired back, creating a booklet to 'explain our side'.

It offered up what would become the company's ultimate line of defence: 'What you may not know is that polystyrene can be recycled.'

*

The person in charge of McDonald's clamshell recycling efforts was a curly-haired, towering 33-year-old named Bob Langert.

Growing up in the 1960s in southside Chicago, Langert had harboured a quiet fascination with men like Martin Luther King and John F. Kennedy. For as long as he could remember, he'd wanted to do something to 'change the world'. He was just never quite sure what that should be.

His moment came in 1988 when his employer, a McDonald's packaging supplier, offered him a 'temporary environmental assignment' with the fast-food chain. 'My job was very specific: to save the polystyrene clamshell by developing a recycling programme for it,' says Langert, who eventually began working directly for McDonald's. 'From day one, I felt like it was my calling to do it because I always cared about the issues of the world.'

By then, the world – or, at least, the slice of it that McDonald's catered to – was increasingly turning against plastic foam. By the late 1980s, Berkeley, California had joined Suffolk County in passing a ban on polystyrene, while governments in Vermont, New York City and Los Angeles had stopped buying foam packaging. Legislatures in 26 US states were considering banning or restricting foam containers.

Recycling was the polystyrene foam industry's key to pushing through escape clauses. As long as companies could *show* that foam was being recycled, legislators indicated they'd allow it. 'What we've found in our lobbying efforts is that focusing on getting recycling markets up and running diverts attention from bans,' Jean Statler, the communications head for the Council for Solid Waste Solutions, told a foodservice magazine in November 1989.

McDonald's began sending packaging waste from 20 of its restaurants to a recycling centre in Brooklyn. Run by polystyrene maker Amoco, the newly opened centre was designed to sort clamshells in the hopes of quelling the backlash. McDonald's said the used clamshells would be reprocessed into resin for new products.

McDonald's also rolled out another far bigger effort, partnering with foam makers Mobil and Genpak in Leominster, Massachusetts, to send containers from 100 New England McDonald's restaurants to a recycling operation the pair named Plastics Again. At a press conference at the Boston Children's Museum, Yastrow said McDonald's would direct customers to put their empty clamshells into new recycling bins.

The polystyrene industry set up a DC-based lobbying group dedicated to defending its plastic. It was headed by Jerry Johnson, a salt-and-pepper-haired former Dow Chemical executive. Soon after taking his new role as head of the Polystyrene Packaging Council, Johnson appeared on *Today*, a popular morning TV show, to debate a third-grader and a fourth-grader. Don Shea watched, paralysed with horror. 'It was like Godzilla versus Bambi,' he says. 'Why would you debate a child?' The *Wall Street Journal* outlined how nine-year-old Bridget Sullivan-Stevens had 'trounced' Johnson, telling him that polystyrene was 'a non-renewable resource'.

Recycling polystyrene was technically possible. But collecting food-soiled containers from restaurants was a different ball game to just sweeping clean cuttings off the factory floor.

Langert cringes as he remembers visiting the plants. Stinking garbage trucks ostensibly delivering empty clamshells unloaded rotting food and used paper napkins. 'By the time it got transported to the recycling centre, it smelled like high heaven and was very attractive to vermin,' he tells me. In his own book about McDonald's, published four years after he left the company in 2015, Langert described the recycling operation as 'a fool's errand, little more than

Bob Langert (*left*) stands alongside the operator of a McDonald's restaurant experimenting with polystyrene recycling in 1990. Image courtesy of Bob Langert

an expensive multi-million dollar PR manoeuvre'. He added: 'I vowed then never to stoop to such greenwashing again.'

Seeing the old McDonald's containers covered in ketchup on a visit to the Amoco plant also helped convince the Suffolk County executive on whom the fate of Englebright's plastic ban hinged to sign that bill into law. 'I left there even more sceptical than when I arrived,' Patrick Halpin recalls. 'There was no way this was going to work at scale.'

Asking consumers at fast-food restaurants to fastidiously separate their trash and empty their food waste didn't work out as McDonald's had hoped. The company tracked how diners behaved and estimated that, at best, a third of customers recycled correctly – most of the time these were mothers with young kids.

'They looked at the instructional signage maybe for a split second, froze, and then just dumped all the trash into the garbage bin,' wrote Langert in his book. 'What we learned from putting a thousand restaurants on this recycling test was that most of McDonald's customers were neither ready, willing, nor motivated to separate their packaging waste after eating a meal.' It was like McDonald's had somehow trained consumers to just dump everything into the general trash in one go, says Langert.

McDonald's had promised that the recycled containers would be turned into plastic to make products like trays and waste baskets. But the resulting quality was so poor that intended customers like Rubbermaid, who had promised to take the recycled plastic for its office and household products, flatly refused to buy it.

Continuing to loudly trumpet the recycling efforts, however disastrous they were in real life, was key to McDonald's fight against CCHW, the schoolkids and the legislators. In 1990, a six-page booklet called *McDonald's and the Environment* began showing up in the chain's outlets across the country. 'Our founder Ray Kroc would pick up litter in the parking lot of his first McDonald's,' it said. 'We base our policies and procedures on the most sound environmental information available.'

The booklet described polystyrene as 'the only foodservice packaging available that is 100% recyclable and is being recycled'. The plastic, it wrote, was 'following in the footsteps of highly recycled materials such as glass and aluminium'. Unsurprisingly, the booklet failed to mention that polystyrene recycling rates for used fast-food containers were close to zero. It didn't tell people that recycling the soiled clamshells, while technically possible, in practice was turning out to be unviable.

The food waste was just one problem. Another was that polystyrene foam is mainly air. Companies paid by material weight didn't make enough to cover collecting the polystyrene and trucking it to plants. It was akin to transporting trucks full of popped popcorn.

Yastrow hired an expensive PR consultant known for helping companies neutralise criticism from environmentalists. Charles Yulish advised McDonald's to make sweeping recycling pledges and form highly publicised alliances with reputable non-profits. 'He said, "*Show* you're green. You have to be proactive,"' recalls Yastrow.

Yastrow took the advice, publicly pledging that, before the turn of the century, there would be a recycled McDonald's restaurant with countertops, seats and flooring made from recycled plastics, and that all the chain's restaurants would recycle their polystyrene by 1992. Needless to say, this did not happen.

McDonald's also pledged to spend an additional $100 million a year buying recycled material, a promise that earned it a medal from President George H.W. Bush. Yastrow still has the medal, as well as the congratulatory letter he got from Bush dated 17th August, 1992. 'Dear Mr Yastrow,' it begins. 'Many thanks for helping to promote the purchase of products made from recycled materials.'

In August 1990, McDonald's teamed up with the non-profit Environmental Defense Fund in a wide-ranging waste-reduction partnership. 'These alliances are helpful,' Yastrow tells me. 'If you get someone who respects you and they say nice things, that's better than saying it yourself.'

At Charles Yulish's urging, McDonald's also struck an alliance with the World Wildlife Fund, a non-profit that was increasingly receptive to corporate donations. McDonald's paid $1 million to fund a magazine for schoolchildren, featuring the WWF logo, that plugged the fast-food chain's environmental activities and defended its use of polystyrene foam. It was the largest-ever educational programme for schools at the time, reaching five million teachers and students.

It wasn't just McDonald's who had cottoned on to the idea that manipulating children's attitudes made good business sense. NutraSweet, a maker of artificial sweeteners, advertised itself as a way to lose weight, while a so-called 'nutrition worksheet' from Chef Boyardee – a maker of ultra-processed canned pasta – listed recipes requiring Chef Boyardee products, but offered nothing in the way of nutritional information.

One agency that created materials for companies targeting schoolchildren described the advantages of the opportunity: 'Early brand loyalty. New sources of business. Profitable secondary markets. A positive corporate image.' Consumer watchdogs complained that students were being subjected to marketing messages in schools. But McDonald's said the magazine 'introduces students to the importance of individual environmental action'.

Emphasising the responsibility of individuals was a tactic increasingly used by packaging makers and consumer goods companies alike. Telling consumers to recycle or not litter was an easy and inexpensive marketing strategy. It allowed companies to appear responsible, without needing to lift a finger to redesign their products to make them easier to economically recycle or reuse. They could also keep pushing the costs of waste disposal and recycling onto cash-strapped municipal governments rather than footing the bill themselves.

Bob Langert said it first, but at least some of what McDonald's was doing back then would now be called greenwashing. The term gained popularity in April 1990 ahead of the 20th anniversary of Earth Day. Greenpeace described greenwashing as companies 'shrouding themselves in cloaks of green to distract the public from their true environmental record'.

Among the early examples of this were Chevron Oil's ads showing off its wildlife restoration efforts; Greenpeace pointed out that the oil company's own operations had destroyed the animals' habitats in the first place and that Chevron was required by federal law to mitigate the damage. Another example was Champion International's sponsorship of boat races when its paper plant spewed millions of gallons of dangerous effluent into a North Carolina river every day.

Ed Rensi, announcing the new school programme, unctuously said McDonald's aim was 'to raise environmental awareness so that all of us celebrate the Earth every day of our lives'. Langert, with the benefit of hindsight, is more candid. 'We were creating all these programmes and policies because now we were at the forefront of being characterised as being bad for the environment,' he says. 'We didn't want to be seen that way. We were desperate to find a way to get the credibility to get this monkey off our back.'

*

It was Halloween of 1990 and Langert was pacing back and forth by an enormous window that looked out over the two artificial lakes flanking McDonald's headquarters in Oak Brook. The building he was in was made from natural materials, such as wood and stone, and designed not to reach any higher than the surrounding trees, some of which were 150 years old. It was conceived to make employees feel immersed in nature and at peace. Right then, Langert felt anything but.

He had just stumbled across a draft press release signed by McDonald's recently named vice president of environmental affairs, Mike Roberts. It announced that McDonald's polystyrene recycling programme would roll out to all the company's 8,500 restaurants nationwide. It was slated to go out the very next day.

A year after the Leominster plant had opened, about two-thirds of what was coming in from restaurants was food and other non-polystyrene waste. By now, Langert was firmly convinced there was no way McDonald's could make recycling work. 'The testing was going horribly,' he says. 'I was shocked that they would do this.'

Only six months earlier, Langert's life had been thrown into disarray by the gory murder of his 30-year-old brother and his brother's pregnant wife. In the months that followed as Langert grieved, he also became determined to live more authentically and make a real difference to the world where he could.

Anxiety mounting, he strode down the hall to Shelby Yastrow's office. The door was open and he could see the general counsel reclining in his leather swivel chair. Yastrow took one look at Langert's face and beckoned him in. Langert laid out what the press release he had chanced upon would say. He told Yastrow why McDonald's couldn't justify even the two recycling operations it was involved in, let alone expanding the programme further.

Yastrow had harboured his own reservations about recycling polystyrene for months. Listening to Langert's panicked plea, he felt firmly that McDonald's should abandon the clamshell altogether. 'We could fight that plastic battle for 1,000 years and bring out every piece of favourable evidence in the world, but a big chunk of the population would never believe us,' he says. 'Plastic was bad in their minds.'

The big hurdle that remained was Rensi. Convincing the McDonald's president had now become a matter of urgency. It had to be done that day, before the press release was sent out.

Over the previous several months, Yastrow had developed a close relationship with the Environmental Defense Fund. Once operating under the motto 'sue the bastards', EDF had greatly softened its stance under the leadership of a straight-talking young lawyer called Fred Krupp. EDF now took the view that working *with* companies, rather than *against* them, was the best way to spark change, making it an outlier in the non-profit world.

As the backlash against McDonald's had swelled and Yastrow had looked for new ways to burnish McDonald's environmental credentials, he had asked Krupp to fly down to Oak Brook to meet him. The two lawyers had hit it off.

McDonald's began implementing several of EDF's suggestions to cut waste, in the process making genuinely beneficial changes. It cut packaging by switching from ready-to-serve orange juice to frozen concentrate, pumped soft-drink mixes directly from delivery trucks into tanks rather than delivering them in cardboard boxes, and packed more French fries into the same case.

Unlike the other environmentalists, EDF stayed away from pressuring McDonald's to switch to reusables. The company's deep-seated antipathy to reusables was evident in a formal written agreement that Yastrow had Krupp sign. It included a clause saying that, if the pair were to work together, EDF couldn't push McDonald's to turn 'fancy' and use real plates, cups and cutlery that could be washed, dried and reused. 'You aren't going to turn us into the Four Seasons,' Yastrow told Krupp.

McDonald's saw switching to reusables – even just for people dining in – as undermining the very foundations of its cheap, quick, carefree model. It wasn't just the logistical hurdles: reusables threatened to mess with the very psychology that underpinned fast food. About 70 per cent of fast-food purchases were made on impulse. The appeal lay not just in the food's affordability and taste, but also in the freedom customers felt in being able to eat half a burger sitting in a McDonald's and the other half as they wandered down the street.

Months into the collaboration with EDF, Yastrow hopped on weekly calls with the non-profit's staff to discuss everything from downsizing napkins to using recycled paper for bags. But so far Krupp, like everyone else, had been unsuccessful in convincing McDonald's to ditch polystyrene. The non-profit had avoided focusing on the issue for fear of derailing the alliance.

That day, Yastrow phoned Krupp to alert him to the impending recycling announcement. Was there anything Krupp could think of that might sway Rensi? Krupp got EDF's lead toxicologist on a conference call with Yastrow and Rensi. She explained how the International Agency for Research on Cancer had recently classified styrene as 'possibly carcinogenic'.[4] There was some evidence

---

[4] In 2019, the International Agency for Research on Cancer (IARC) reclassified styrene as 'probably carcinogenic to humans'.

showing that styrene oxide was cancerous in animals, which indicated it might have a similar impact on human beings. The toxicologist minced no words in explaining that she thought recycling
polystyrene was a very bad idea.

For Rensi, after three years of staunchly defending his precious
container, the allegation that it could cause cancer went too far.
'That was the straw that broke the camel's back,' says Langert.
'Rensi very reluctantly pulled the plug on it. It's not like he wanted
to do it.'

On the morning of 1st November, 1990, McDonald's PR
department did send out a press release, but not the one Langert
had found.

Instead, in a shock move, it announced it would abandon the
clamshell.

Rather than admitting defeat on environmental or even health
grounds, Rensi cited an emotional reason for why McDonald's
would stop using foam. 'Although some scientific studies indicate
that foam packaging is environmentally sound, our customers just
don't feel good about it,' he told reporters. 'So we're changing.'
Later that day, in a TV interview, Rensi softened slightly, saying
his decision was at least partly environmentally motivated. 'We did
it for the good earth, we did it for our customers and we did it for
ourselves,' he said piously.

Thirty-five years after the clamshell controversy, Rensi isn't
eager to talk about it, declining my interview request via an email in
which he cc'd his lawyer. 'I have no interest to [sic] discussing this
matter,' he wrote. 'I retired from McDs nearly 30 years ago. Call
their CEO and discuss with him.'

Getting rid of the clamshell also meant the demise of the McDLT.
By April 1991, it was history. For its other food, McDonald's
swapped the clamshell for waffle-patterned paper sandwiched on
either side of a sheet of low-density polyethylene. In other words, it

was plastic layered with paper. The wrap's quilt-like effect created air spaces in the paper that allowed it to retain heat.

'It's a real breakthrough,' said Rensi.

*

In switching to paper – the material it had abandoned 15 years earlier partly because of concerns about deforestation – McDonald's scored a huge public relations win. 'McGreen Inc.' was the headline from the *Washington Times*. 'Let's Hope More Companies Copy McDonald's Effort', enthused the *Seattle Post-Intelligencer*. 'McDonald's is at last showing some McSense on the environment,' wrote a *New York Times* reporter whose story was titled 'Greening of the Golden Arch'.

Burger King took out snarky full-page ads in seven newspapers mocking McDonald's for its copycat move. 'Welcome to the club,' said its ad. 'We wonder what the planet would be like if you had joined us in 1955.'

But swapping one single-use material for another made for a dubious environmental gain.

As McDonald's alluded to in 1975 when it decided to use plastic, paper can cause deforestation and hurt biodiversity. Making one A4 sheet of paper requires around 10 litres of water on average. Paper production uses lots of different chemicals, including some which researchers have flagged as being potentially harmful to human health. In time, the paper food packaging used by many fast-food companies, including McDonald's, was found to contain PFAs (per-and poly-fluoroalkyl substances), the most commonly used of which have been linked to immune system suppression, higher cancer risk and liver problems.

After pressure from environmentalists and lawsuits from consumers, in 2021 McDonald's said it would stop using purposely added fluorinated compounds in most markets by 2025. It declined to tell me what it was replacing them with. Researchers have raised concerns that replacements to PFAs could be similarly harmful.

Although McDonald's was careful not to talk about it, the paper package it switched to in the early 1990s was very hard to recycle, given its layer of plastic was difficult to separate. The wrap wasn't even made from recycled material since regulations in the US, as in many countries, don't allow food packaging to be made from recycled paper for safety reasons. Today, recycling paper lined with plastic and other coatings is still difficult, while the regulatory requirement that new food packaging must be made from virgin paper remains in place.

In time, McDonald's switched to a cardboard box that can technically be recycled. But it's still grappling with the same issues it faced in the late 1980s: convincing fast-food customers to stop and scrape food waste out of packaging so it can be recycled is really hard. Leftover food is even more damaging to paper than it is to plastic because it can release oils and fat, reducing the fibre's strength.

McDonald's continued to use polystyrene foam cups for nearly two decades after abandoning the clamshell. Under pressure from non-profits, it replaced these with plastic-lined paper cups in 2018. The cups must be collected separately and sent to specialist mills for recycling since the plastic liner is hard to separate. Cold 'paper' cups for soft drinks and shakes are not recyclable at all because they're lined with plastic both inside and outside, to protect against condensation. The plastic stops the water that's used at recycling plants to break down paper from accessing the cup. Paper packaging that goes to landfills can remain intact for decades. If it does break down, it releases methane, a potent greenhouse gas, much of which isn't captured.

Back in the 1990s, several environmentalists flagged that McDonald's, in clinging to disposability, may have swapped one problem for another. But many celebrated the chain's abandonment of foam food packaging as the start of a new chapter. 'McDonald's has scored an environmental touchdown,' said Krupp from the Environmental Defense Fund. 'This action is the beginning of a new era of environmental problem solving.'

McDonald's abandonment of foam clamshells rocked the plastics industry to its core. The chain alone had accounted for 8 per cent of the 1 billion pounds of foam packaging sold in the US each year. Others followed suit and polystyrene foam sales fell as much as 15 per cent. 'Everybody is going to be re-examining their rationale for using polystyrene,' predicted Jerry Johnson of the Polystyrene Packaging Council. 'It's a big blast, symbolically.' Already, through 1990, around 70 different bans on polystyrene had been proposed across the US.

'It was a giant wake-up call,' says Don Shea of the Council for Solid Waste Solutions, who by now was knee-deep in his attempts to ward off bans of other plastic products. 'What were we doing?' he continues. 'Probably a lot of praying. We were kind of like the Dutch boys and girls putting our fingers in the dike to keep the dam from bursting.'

CHAPTER 4

TALKING SHIT

The purpose 'was to make people feel good about disposable products'

Scott Stewart was three days into his new job as a public relations manager for Procter & Gamble in October 1988 when his boss swept into his office to say she was scheduling a crisis meeting for that afternoon.

Word was that the National Association of Diaper Services – or NADS, the trade body for companies that washed and delivered cloth diapers – was on the brink of publishing a study showing that the 18 billion disposable diapers thrown away each year made up 2 per cent of America's landfills by weight. That ranked them as the largest consumer product in landfills after newspapers and drinks containers. Worse still, the study, written by a guy called Carl Lehrburger who nobody at P&G had ever heard of, would say that taxpayers were being billed $300 million to landfill and burn disposable diapers each year.

'I didn't know then that this was going to define my life for the next five years,' Stewart says.

In the months after the *Mobro* made headlines, disposable diapers came under attack just as McDonald's clamshells had. Many saw them as representative of a throwaway culture created by companies focused on making the highest possible profits by offering short-term benefits and convenience with little thought for what the future would look like.

The diapers were bulky, smelly and increasingly littered. It was easy to imagine them filling up landfills. Some environmentalists estimated that each disposable diaper, worn for a few hours at most, would last 500 years in the average landfill.

The diapers had also only caught on relatively recently; most parents using them for their babies in the late 1980s had started life with their own bottoms swathed in cloth. The threat of a proven alternative had always loomed large for P&G, which had poured millions into ensuring parents turned their backs on cloth.

At six foot five with a bristling moustache, Stewart cut a figure that was hard to miss. He wasn't exactly new to P&G – he had spent the past four years as a sales rep in Kentucky – but this was his first job at corporate headquarters. Reclining in his small windowless office on the second floor of the Sycamore Building in downtown Cincinnati, flanked by a filing cabinet and a modest desk, Stewart, then 34 years old, felt he had finally arrived. 'I was the dog that caught the car,' he remembers.

Stewart had grown up in the suburbs of Louisville, Kentucky, next door to a Catholic grade school. On sunny days he would ride his Stingray bike across the river to downtown New Albany. He studied economics and public policy at Indiana University and worked as an office director for two senators, but eventually decided it was time to move to the private sector. 'Frankly, I wanted a job that would pay health insurance and give me a company car,' he says. When, in 1984, he saw an ad in the Louisville *Courier Journal* for a P&G sales job, he applied straight away.

At first, Stewart visited hospitals to convince administrators that P&G's disposable gowns and sheets were better than the washable linens they were using. Later, he shifted to telling nursing homes why they should move from using washable cloth diapers for incontinent residents to P&G's disposable ones. 'The imprimatur of P&G overwhelmed any embarrassment about what I was doing day in and day out,' he says. Amiable and talkative, he got on well with the nursing home staff, often spending as much time commiserating over their difficulties with recalcitrant residents as he did selling P&G's diapers.

Just as fast-food chains had switched away from reusable dishware and cutlery, the shift by hospitals and nursing homes to disposable 'non-wovens' – which were made from common plastics but mimicked cloth – had accelerated in the late 1960s as labour costs rose. By 1968, the non-wovens industry was large enough to need its own trade body. It chose a refreshingly candid name: The Disposables Association.

Back in the early 1980s, few people were worrying about plastic waste. Once Stewart had convinced his customers that disposables weren't more expensive than reusables, he met little resistance. He studied his sales talking points closely enough that, four decades later, he can still parrot them with no hesitation. Reusable cloth briefs with plastic pads rather than P&G's fully disposable diapers 'had the potential for skin to break down and cause harm to patients', he says seriously. Stewart was – and remains – fiercely loyal to P&G. 'I would fall on my sword for that company.'

On the afternoon of 13th October, 1988, when Stewart sidled into the meeting that his boss, Sue Hale, had called, seven other faces in the room – most belonging to people he hadn't met – turned to watch him enter. 'At that point in time, I really didn't understand the depth of what people who did what I did did,' Stewart says. 'My learning curve was vertical.'

Listening to Hale speak about the battle P&G inevitably had ahead to defend its most profitable product, Stewart felt he had arrived at the first major moment of his career. 'This is the first time I walk in and see serious people dealing with a serious issue,' he says. 'That's where my career as a crisis communicator was born.'

*

Balding and bearded with large spectacles, 37-year-old Carl Lehrburger, author of the diaper study, wasn't someone you'd shoot a second glance at if you passed him on the street. One former colleague describes him as 'an absent-minded professor type, very smart but could forget to tie his shoelaces if someone didn't remind him'.

Still, Lehrburger had oodles of determination and a passion for the planet that was hard to beat. 'This has to be seen in the context of the war on nature, the most existential battle we face on planet earth,' he says when I interview him, worried that a book chapter on diapers will miss the bigger point. 'The diaper war was a small aspect of the disposable waste economy and the many other environmental issues that are still in play. Failure to put the diaper issue in context of a corporate war against nature affecting future generations will likely trivialise that particular skirmish.'

P&G's executives viewed him with a mixture of bemusement and trepidation. 'Carl looked like the person you'd have from Central Casting to do this,' cracks Stewart.

Lehrburger lived in Sheffield, a town nestled in the floodplain of the Housatonic River in Massachusetts. He had first decided to study how much waste disposable diapers were generating after stumbling across an article in which the industry argued that the diapers made up just 0.1 per cent of the US's municipal waste. 'It was greenwashing,' he says firmly. 'They were presumably counting dry diapers. I took them to task, so to speak.'

Lehrburger's environmental activism had taken root in the 1970s at the University of Colorado in Boulder, when he helped organise a resistance movement against nuclear energy. Subsequent jobs after graduation included making solar greenhouses, working in recycling and turning waste wood into pellets for fuel.

Eventually, Lehrburger found himself working part-time at a fish smokehouse, arriving at work before 5am to unload fish packed in ice and salt. 'I was salting, marinating, smoking and hand slicing salmon,' he says. 'It was quite a benefit because we ate very well.' When he got home each day, Lehrburger played with his toddler Ethan – who didn't seem to notice that his dad smelled strongly of fish – but found himself feeling restless. On a whim, he decided to turn one of his most disliked chores, changing Ethan's diapers, into a project. He would use his baby's daily discharge to figure out how much diapers were actually contributing to America's solid waste.

And so began two years through which Lehrburger spent much of his time thinking about diapers. He bought a postage scale, complete with an aluminium bowl, and set it up in his country-style kitchen. There he carefully weighed Ethan's diapers – when they were empty, when they contained urine and when they sagged with poop.

A meticulous record keeper who would go on to keep 25 journals in which he documented and analysed more than a thousand of his dreams, Lehrburger recorded the weight of each week of diapers. He also logged cubic measurements so that he could talk about the issue in cubic yards, should anyone ask. 'How you documented this was very controversial,' he says. 'I had been in the solid waste management industry, so I knew that weight was the key measurement because these things had to be transported.' Lehrburger supplemented his own number-crunching with broader data. 'I was able to do some statistical analysis of the weight of the diapers going into a residential trash bin that was then going into a landfill at a certain price so I'd have a database to draw from to write an article.'

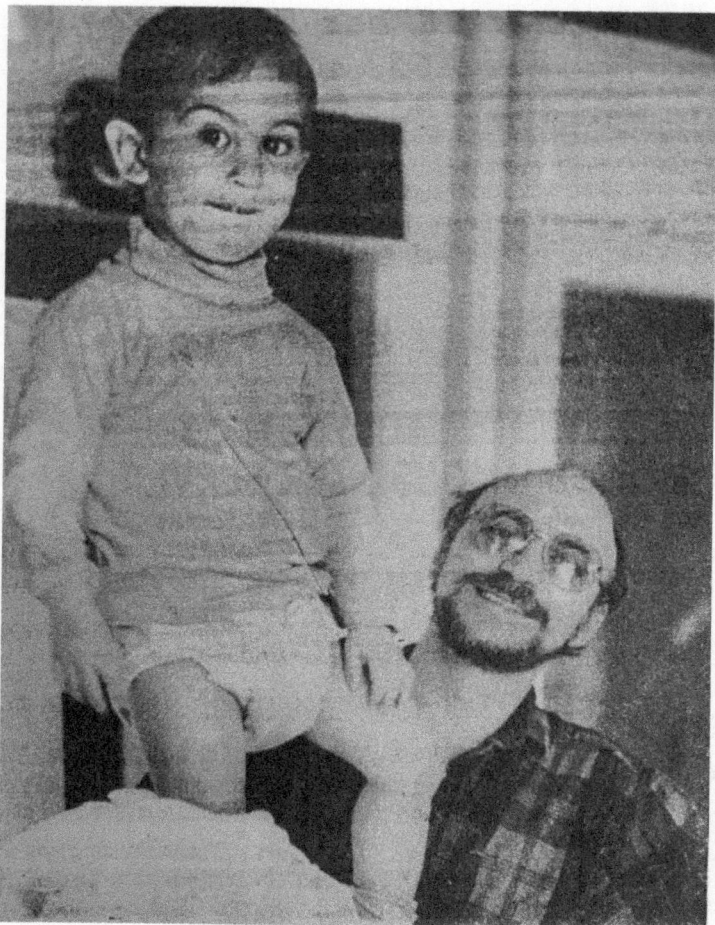

Carl Lehrburger with his diaper-clad son Ethan aged two and a half. Image courtesy of the *Berkshire Eagle*

His article, 'The Disposable Diaper Myth', was published in the fall of 1988 in an obscure magazine called *Whole Earth Review*. In it, Lehrburger laid out his findings with relish. 'Huggies, Pampers and Luvs are not "disposable" at all. We throw about 18 billion of them away each year into trash cans and bags, believing they've gone to some magic place where they will safely disappear,' he wrote. 'The truth is most of the plastic-lined disposables end up in landfills. There they sit, tightly wrapped bundles of urine and faeces that partially and slowly decompose only over many decades.'

Back at the Sycamore Building in October 1988, Sue Hale handed out copies of Lehrburger's article, reading the more colourful bits aloud to the room. 'What started out as a marketer's dream of drier, happier, more comfortable babies has become a solid-waste nightmare of squandered material resources, skyrocketing economics and a growing health hazard,' she read, grimacing.

NADS, the trade body for the cloth diaper makers, heard about Lehrburger's findings and offered him $15,000 to do a more comprehensive study. He hired two researchers and spent the next six months working on it. In the meantime, journalists across the country got wind of Lehrburger's work, giving readers an advance look at the study's juicier findings. Less than 5 per cent of people changing disposable diapers were following manufacturers' recommendations to empty faeces into the toilet before throwing the diapers away. Ten per cent of disposable diapers contained viruses, including live poliovirus from immunisations. For every $1 consumers spent on disposable diapers, 8 cents were spent by taxpayers on their disposal.

Environmentalists used Lehrburger's study to report that the volume of diapers thrown away each year would fill an Islip-style garbage barge every six hours. 'For more than a decade, the disposable diaper has been a symbol of contemporary child care. Today it is emerging as a symbol of the nation's garbage crisis,' proclaimed the *New York Times* in December 1988.

The environmental impact from disposable diapers had been overlooked, warned Lehrburger. States, he said, should now look to minimise or even eliminate disposables. He advocated a return to cloth diapers since washing these meant that faeces and urine would safely be taken care of by the sewerage system and the fabric could be reused 200 times before being recycled.

Lehrburger also raised a question that, nearly four decades later, continues to be hotly debated across the US and many other parts of the world: Who should bear the costs of waste disposal?

'As a society we have been sold on the idea of "disposability" without full recognition of the costs associated with convenience,' he wrote. 'We have now reached a point where we must question not only the issue of whether alleged "convenience" justifies tossing 18 billion single-use diapers into the solid waste stream annually, but whether the responsibility for safe disposal of single-use products rests with the manufacturer or the consumer.'

*

The cloth diaper rental and laundering industry began in 1933 when Albert Lau, a Chicago bond salesman, bought three trucks, hired six workers and, with the help of his wife, opened Dy-Dee Wash. Soon Dy-Dee and rival companies were opening outlets in droves across the country. They supplied 80 to 100 diapers a week for $1.80. Customers also received a container to store soiled diapers, along with deliveries of fresh diapers twice weekly.

The process was efficient and sanitary: the cloth diapers were washed for one hour, put in an extractor for 15 minutes to wring out any additional water and then dried and sterilised in a tumble dryer for half an hour at 180 degrees Fahrenheit. After this, the diapers moved to sorting baskets to be inspected and folded.

By the time the Second World War ended, Dy-Dee was laundering 1.4 million diapers a month for individuals and another 200,000 for hospitals. Like any business, the operation proved expensive at first but, as it scaled, costs declined dramatically. Paying $1.80 a week amounted to 50 cents more than washing one's own cloth diapers at home, a price many parents were happy to pay for the convenience.

The diaper rental companies also added a personal touch. They offered free photo shoots of new babies to parents who signed up. They phoned to solicitously enquire after ill babies. The diaper trucks used horns that loudly but soothingly played 'Rock-a-bye Baby' to alert parents to come to the door with their dirty diapers, not unlike how an ice-cream van alerts neighbourhood kids to its presence. The drivers, many of them fathers, offered advice on soothing crying babies and pinning cloth diapers.

For many young families, the diaper laundry service became as core a part of their morning routine as the milkman who dropped off returnable glass bottles. 'Getting diapers washed commercially has become a part of American thinking,' wrote *Kiplinger* magazine in 1947. 'Psychology is deeply rooted in favour of the diaper service.'

The company that would upset this happy equilibrium was, of course, P&G.

Legend has it that in the mid-1950s, a P&G chemical engineer named Victor Mills was forced to change the cloth diaper of his grandson, Vic, who was named after him. Mills was apparently shocked by how poorly it fit, how messy the process was and how quickly the diaper soaked. Or that's P&G's version of events anyway. Mills's granddaughter Gracie Schild recounts a slightly different version. 'My mother says there's no way my grandfather was actually changing diapers,' says Schild. She surmises that her mother changed little Vic while Mills looked on and tut-tutted.

In 1957, P&G acquired Charmin, a maker of toilet paper, facial tissues and paper napkins. It charged Mills, who headed up P&G's 'exploratory products division', with creating new products within the paper unit.

Mills, no doubt inspired by his grandchildren, decided the time was ripe for a disposable diaper. Disposable diapers had in fact existed in the US since at least the 1940s when Johnson & Johnson launched one called Chux. Individual entrepreneurs came up with their own versions too; Connecticut housewife Marion Donovan coupled a shower curtain with absorbent padding to create a disposable diaper she called the Boater. But they were expensive and by the mid-1950s less than 1 per cent of American babies' bottoms were swathed in disposables. At the time, the cloth diaper services industry paid little heed to such niche attempts. Instead, it saw the advent of the washing machine as its biggest threat.

P&G calculated that there were more than 15 billion diaper changes in the US each year, which made for an enormous untapped market. Mills recruited men to form a research group dedicated to creating a disposable diaper that would succeed where the rest had failed.

For nine months, Mills's team trialled different designs. One researcher brought a Betsy Wetsy doll into P&G's labs, hooking the legs up to a treadmill so it would walk. The doll was designed so that if you poured water into the mouth, it would trickle out the other end. The researchers delighted in testing their various prototypes on it.

Back then, nobody at P&G was thinking about making a diaper that would involve tossing faeces and urine into the trash. Instead, the researchers designed an absorbent, flushable pleated diaper pad to be inserted into reusable plastic pants.

In the late 1950s, P&G's marketers began testing the diapers in Dallas. But in the heat of Texas, the reusable plastic pants created a hothouse effect and the diaper was a flop. 'The kids were just sweating and it was awful,' recounts Schild. Mills sent his team back to the drawing board with orders to do better.

Six months later, the researchers had created a new diaper. This time, it was a single piece and entirely disposable. It had a thinner plastic backsheet that kept the moisture in but didn't generate the heat that plastic pants did. It also kept babies drier by using a porous water-repellent topsheet that sat next to their bottoms. This let fluid pass through but prevented it from coming back.

Mills showed the new design to P&G's vice president of research, Gib Pleasants. The story goes that, at the start of the meeting, Pleasants had planned on killing the whole project. 'I can't find it in my heart to stop it now,' he said after he saw the one-piece version. 'Test that diaper.'

By March 1959, P&G had 37,000 of the diapers hand-made and ready for testing in Rochester, New York. Some of them had tapes, some pins. Parents overwhelmingly preferred the pin-on versions. Two-thirds said the disposables were as good as, or better than, cloth.

P&G finally had a design that worked. The company's marketers began working the phones, surveying mothers for their thoughts on what the new diaper should be called. Tenders, Dri-Wees, Winks, Tads, Solos, Zephyrs and Pampers were among the contenders. Overwhelmingly, mothers said they liked Pampers, which evoked feelings of tenderness and care. Mills liked the name too. 'He thought it's as much about pampering the parent as the baby,' says Schild.

In the meantime, P&G's engineers were struggling to design a production line that could take the diaper's three layers, speedily fold them in a zigzag fashion and glue them together. The glue applicators dripped heavily and the absorbent wadding generated so much dust that the machinery could only run for a few minutes before needing to be cleaned. One engineer described it as 'the most complex production operation the company had ever faced'.

Eventually, P&G's engineers triumphed. In December 1961, the company began its first major test of the new Pampers, in Peoria, Illinois. For consumer goods makers, Peoria represented the typical medium-sized US city – 'Will it play in Peoria?' had long been a figure of speech across the industry. How customers in Peoria reacted would provide the litmus test for whether Pampers could succeed nationwide.

But, within six months, it was clear that the new diaper was also a flop. At the time, parents paid about 3.5 cents apiece for diapers from a diaper service, while those who washed diapers at home paid less than half that. At 10 cents apiece, Pampers was far too expensive.

P&G had spent its largest amount ever on new equipment in fiscal 1962 and the company's budget was already stretched. But Neil McElroy, P&G's chairman, saw enormous potential in the disposable diaper. He decided to more than double production, gambling that the efficiencies gained would allow P&G to slash the price of Pampers to 6 cents.

At the next test, this time in Sacramento, California, the newly affordable Pampers flew off shelves.

As P&G rolled Pampers out more widely, people began buying the disposable diapers for regular wear. Day-care centres that had once used cloth diaper services now began requiring parents to send babies in wearing disposables. Just a few years after Pampers launched, 25 per cent of diapers being changed in the US were disposable, up from 1 per cent before the brand was created.

The success went well beyond what Mills and others at P&G had envisioned. 'They never in their wildest dreams thought this would become a daily use item – it was for travel, hospitals, sending the kids to Grandma,' says Schild. 'It was never conceived as something that would be affordable for everyday.'

Although the diapers were largely made from cellulose – which provided absorbency – and hence housed under P&G's paper division, plastic was a key component. The backsheet was made of polyethylene while the topsheet was polypropylene. Over the years,

Victor Mills, inventor of Pampers. Image courtesy of *Arizona Daily Star*

as P&G slimmed down the diaper, the proportion of plastic to cellulose would grow.

The fully disposable design turned out to be absolutely crucial to Pampers' success. One of P&G's rivals, Scott Paper, made a two-piece diaper by putting a flushable pad into reusable plastic pants. But it leaked and the pad, when flushed, clogged toilets. BabyScott diapers went down so badly with consumers – some of whom sent the company their plumbing bills – that analysts dubbed the foray 'Scott's Vietnam'. In 1971, Scott Paper killed the diaper, taking a $12.8 million write-off.

Disposable diapers promised enormous new sales not only for P&G but also for the makers of paper and plastics – if only they could convince parents to stop using pesky reusable products.

In 1970, International Paper Company took out a pricey two-page magazine advertisement entitled 'The Story of the Disposable Environment'. It began: 'Everything the baby wears or touches – virtually the entire environment in which he lives – can be disposable.' The company's list of targets included diapers, bedding, shirts and pants. Parents, it promised, would avoid mountains of work, diaper rash and contaminating their washing machines. International Paper's view of disposability even stretched to nursery furniture, which it said could be 'colourful and sturdy yet it will cost so little you'll throw it away when he outgrows it'. The ad wrapped up by saying: 'The disposable environment – the kind of fresh thinking we bring to every problem. Nice to know it's at your disposal, isn't it?'

P&G, which was America's largest advertiser, was also pouring big money into convincing parents that cloth diapers were ineffective and unfashionable. In 1970 alone, P&G spent $4.9 million to advertise Pampers. Night after night, TV ads extolled the convenience of Pampers to housewives up and down the country. Side-by-side comparisons showed a leaking cloth diaper held under a tap alongside plastic-lined Pampers, which P&G claimed leaked nothing.

'Pampers – the discovery that made diapers old fashioned!' was the company's bold tagline for its blockbuster new brand.

*

By the early 1970s, disposable diapers covered 37 per cent of American babies' bottoms and Pampers had become P&G's second-largest brand. The company had a 90 per cent hold over the US's $200 million disposable diaper market. It was also eyeing the huge potential overseas and, in 1973, it began testing Pampers in Germany. But Pampers' dominance wasn't to stay unchallenged for long.

In 1971, Darwin E. Smith, a thickly bespectacled, Harvard-trained lawyer, became the CEO of Neenah, Wisconsin-based Kimberly-Clark and set his sights on dethroning Pampers. Since 1968, Kimberly-Clark had sold Kimbies. The diaper brand did well enough but leaked often and was no match for the heavily advertised Pampers. By the mid-1970s, many inside Kimberly-Clark thought the company should pull out of diapers altogether.

Not Smith. He was convinced that America was on the cusp of a baby boom and saw an enormous runway for growth amid the culture of disposability sweeping the country. Pampers had already laid the groundwork to convince millions of parents that switching to disposables from cloth was a no-brainer. Kimberly-Clark just needed to ride this wave.

In 1978, Kimberly-Clark launched Kleenex Huggies. By now, P&G had two diaper brands – Pampers and Luvs. If P&G's diapers made babies feel pampered and loved, Smith hoped that Huggies would remind them of being wrapped in a soothing embrace.

A Kimberly-Clark engineer found a way to stop runny poop from leaking out of diapers, a problem parents were increasingly grappling with as doctors encouraged mothers to breastfeed for longer. He added stand-up elasticised flaps and elasticised leg openings.

Early Huggies ads from the 1980s showed baby stockbrokers, professors and advertising executives, all unable to do their jobs because they were wearing leaky diapers. Being put in Huggies pulled them out of the doldrums. 'Basically, the message – whether directly or indirectly – was that Huggies babies were happy babies,' says Wayne Sanders, who succeeded Smith as Kimberly-Clark's CEO. 'That's it. They stop leaks. Leaks were a big deal.'

Huggies got a major boost in 1983 when Kimberly-Clark launched diapers with re-fastenable tabs. The popularity of the new Huggies was particularly galling to P&G, which had begun testing the tabs first but hadn't moved quickly enough to get them out. 'At that point, P&G was very, very conservative about product upgrades,' remembers Terry Batsch, who started at P&G in the late 1960s and worked in diaper product development for over 30 years. 'We were probably slower than we needed to be.'

Huggies grabbed market share from Pampers and Luvs, costing P&G up to $30 million a year in lost sales. Smith also poached many of his Cincinnati rival's best marketing and research people by offering fat salaries and opening a shiny new operating headquarters north of Atlanta.

He deployed squadrons of marketing executives to track down pregnant women and new mothers, giving them Huggies-branded baby-care booklets in childbirth classes and free Huggies diapers in hospitals. Finding pregnant women wasn't hard: maternity clothing stores and birthing classes were selling customers' names, addresses and due dates.

Research showed that targeting mothers early had a high payoff since they often continued with the diaper brand they encountered first. 'Diapers is a very complex, competitive, technology-driven category,' says Sanders, 'because there are about 11,000 babies born every day, which is really good news. But the bad news is there are 11,000 customers leaving the diaper category each day. You have to win parents over each time and there are very few categories where you constantly have to do this.'

Kimberly-Clark spent $35 million to create its own database, identifying 75 per cent of the new mothers in the US. It included not just their names, but their demographics and a history of what kind of products they liked. Kimberly-Clark also bought subscribers' information from *American Baby* – the most trusted source of information for new parents.

By the end of 1985, Huggies had leapfrogged Pampers to become America's biggest diaper brand.

Inside P&G, tensions were running high. By then, diapers were P&G's most important product, bringing in close to a quarter of all its profits. The company needed something revolutionary to stop Huggies from continuing to erode its market share.

*

Terry Batsch, the P&G diaper researcher, had been mulling over ways to make Pampers better when he was struck by the notion that perhaps P&G shouldn't be taking what parents said at face value. When P&G conducted surveys, American parents consistently claimed they wanted thicker diapers. What if, wondered Batsch, what they actually wanted was a more absorbent diaper?

Back then, Pampers' absorbent core was made of fluff pulp from softwood. The more fluff, the more 'voids' for urine to be absorbed. Diaper makers were careful to compress the fluff only up to a certain point, since further compression was thought to decrease the diaper's capacity to hold urine. This meant the thicker the diaper, the more absorbent it was.

'I postulated that if we could design a more underwear-like thin diaper that outperformed the current thicker diapers, moms would enthusiastically prefer it,' recalls Batsch. He asked his team in P&G's Winton Hill Technical Center in Cincinnati to make four hand-made diapers: one at the current thickness of Pampers, one that was three-quarters as thick, one half as thick and one a quarter as thick.

Batsch carried the designs around the technical centre, stopping by the desks of 12 women. 'We called them all secretaries at the time,' he tells me. 'Which of these would you choose if I could guarantee they all have the same absorbency?' he asked the women (who were not, in fact, all secretaries). Eleven of the 12 chose the thinnest version; the one who didn't told Batsch she didn't believe a thinner diaper could have the same absorbency as a thicker one.

'After that exercise, I was convinced of two things,' says Batsch. 'One, my theory about preference for a thinner design was correct; and, two, we had a considerable challenge if we wanted to convince moms the new design would work.'

P&G's scientists had for years looked for ways to make diapers more absorbent. One promising technology involved using superabsorbent polymers made by chemical companies like Dow. The scientists so far had been stymied by gel blocking, which is when particles of superabsorbent material swell to form a dam which stops fluid from passing through into the rest of the absorbent structure.

A pair of P&G scientists solved this issue when they discovered that gel blocking didn't happen if the voids between swollen particles were fixed in place. Once wetted, the voids stayed in position, allowing the fluid to get to the superabsorbers and the entire diaper to swell.

The result was Ultra Pampers: a new superabsorbent diaper, rammed with synthetic crystals. It absorbed up to 80 times its weight in liquid, which turned into a gel that locked away moisture. P&G deployed the technology in Japan, where small apartments meant parents had little appetite to store bulky diapers. Batsch began working with P&G's marketers on how best to convince American parents that thinner diapers could be more absorbent than thicker ones.

One ad involved the 'wring-out demo', showing a cloth diaper and the new, more absorbent, Pampers side by side. 'We poured water into both of them and we wrung 'em out,' says Batsch.

'Wouldn't you know it, water ran right out of the cloth diaper, but you wrung Pampers and nothing came out.'

To launch Ultra Pampers, the company pulled together a battle plan, starting with a press conference at a New York hospital. The venue was carefully chosen so that P&G could emphasise the endorsements Ultra Pampers had received from paediatricians. To counter concerns that the gel was so absorbent it might extract fluid from the body, making babies more susceptible to infection, Batsch and a colleague performed a mini circus act. 'He ate them,' says Batsch, explaining how his colleague would pop the super slurper crystals in his mouth at meetings with doctors and salespeople.

P&G's hard work paid off. In just seven months, Ultra Pampers captured 15 per cent of America's disposable diaper market. Wall Street analysts predicted they'd capture 75 per cent of the market within a year. 'These diapers are the biggest success of any new format of an existing product in the history of consumer packaged goods,' enthused one Wall Street analyst who followed the sector.

Kimberly-Clark wasn't about to roll over. Within a few months, the company swept the market with its own slim, superabsorbent version of Huggies. P&G sued Kimberly-Clark for patent infringement, but lost after a judge ruled that P&G's patent couldn't be enforced.

The years that followed saw a dizzying array of new diaper launches, accompanied by huge marketing blitzes, as companies jockeyed for dominance. Diapers now came in blue with extra padding in the front for boys and pink with extra padding in the middle for girls. They featured *Sesame Street* muppets and Disney characters. They came in pastel shades for spring and bolder colours for summer. Supermarkets got in on the act too, making disposable diapers with names like 'Designer Denim Diapers', fashioned to make babies look as if they were wearing jeans.

The diaper market exploded. By 1987, disposable diaper sales in the US stood at $3.3 billion, up from $90 million in 1969. Before the end of the decade, disposables would account for 75 per cent of all diaper changes in the country. 'In all likelihood, disposable baby diapers represent the most dynamic new product category in consumer marketing history,' wrote a marketing columnist for *Ad Week* in 1988.

Cloth diaper makers found it hard to fight back. 'Procter & Gamble spends more money on advertising than we generate in sales,' a manager at cloth diaper services company Dy-Dee told the *Chicago Tribune* despairingly.

Rather than limit their ambitions to babies, the disposable diaper giants set their sights higher (quite literally). In 1989, Kimberly-Clark launched training pants called Huggies Pull-Ups. Suddenly, millions of toddlers who might have gone from diapers to regular underwear were spending another year in disposables. 'We had kids singing a jingle, "I'm a big kid now",' says Wayne Sanders. 'That thing was so powerful. Pull-Ups took off like a rocket.'

The technology that allowed for new, thinner diapers also paved the way for P&G and Kimberly-Clark to double down on selling diapers to adults. With 10 million American adults estimated to have incontinence, the potential market was enormous. In 1988, one maker of adult disposable diapers estimated that, by the year 2000, the market could be worth $6 billion – double the current baby diaper market.

*

Within a decade of launching Pampers, P&G was ploughing millions of dollars into ad campaigns to convince parents in 70 countries to buy its disposables.

In colder countries, children had traditionally been diapered in cloth, while in warmer regions, many hadn't worn diapers at all. In China, children's clothing came with a slit in the back. Called *kaidangku*, the slit trousers allowed its wearer to urinate and defecate while on the go. Parents were wise to the signs of when children needed to empty their bowels, whisking them off to sit over a container. Potty training started as young as six months, with any messes just cleaned up. By the age of one, many children indicated when they needed the toilet and had control of their bladders.

To convince the Chinese to embrace disposables, P&G launched a wholesale marketing assault, promising that disposable diapers meant better sleep and hence better brain development. Its efforts included erecting a 660-square-metre photomontage in Shanghai featuring crowd-sourced photos of sleeping babies. To much fanfare, it also handed out 'Golden Sleep Ambassador Certificates' to thousands more babies, displaying photos of their well-rested 'morning smiles' outside a major sports centre in Beijing. Fast forward a few years and many Chinese parents were spurning kaidangku, which they now described as 'uncivilised'.

Many of the countries P&G was targeting had no organised garbage collection service. In countries that did, like the US, billions of diapers were suddenly going to landfills, where they would remain for hundreds of years. In the wake of the *Mobro*, it became clear that P&G, for the first time, would need to find a way to manage that environmental impact. Or at least the public's perceptions about it.

'Not since the automobile begat the suburb has a product so essential to the American lifestyle become so maligned as a symbol of pollution and waste,' wrote the *New York Times* in September 1990, some two years into covering the diaper issue. 'The disposable diaper, which during the 1970s and '80s became as indispensable to working parents as *Sesame Street* and strained carrots, is suddenly the environmental bad boy of the 90s.'

Edwin Artzt took over as P&G's CEO just as the reputation of disposable diapers plummeted to an all-time low. A poll published in the *Wall Street Journal* on 20th April, 1990 showed that a staggering 74 per cent of Americans favoured banning disposable diapers to reduce waste.

Around 20 US states were considering proposals ranging from taxing disposable diapers to all-out bans. Nebraska had already approved a ban on disposable diapers that didn't degrade, slated to take effect in 1993. Legislators in Maine had passed a law requiring day-care centres to accept children in cloth diapers. New York was considering requiring hospitals to provide new mothers with brochures warning of the environmental threat posed by disposables.

'Until that point, life in that world was, how do you build the business? How do you compete?' says Scott Stewart, the PR executive. 'Suddenly you have an outside play here. Artzt made it very clear that he wanted the company to be proactive on this. Our objective was to protect the business.'

By the early 1990s, nobody at P&G was underestimating how serious a threat anti-disposables sentiment posed. Unlike seemingly far-away issues like deforestation or air pollution, consumers had it within their power to make big changes when it came to diapers. An alternative to disposables existed and already many had begun switching to it.

'There was a lot of environmental advocacy for cloth diapers and the business was really growing,' recalls Tim Sergeant, who in 1990 worked as a merchandising manager at Gerber Childrenswear, the largest American maker of cloth diapers. 'We were really trying to use what was already going on to our advantage, and it was working.'

Inside P&G, which sold everything from orange juice to laundry detergent in disposable containers, executives worried that the diaper backlash could metastasise into something much bigger. The most glaringly obvious problem was that while disposable diapers made up just 2 per cent of America's solid waste, everything else that was being thrown away in larger quantities – yard waste, food waste, newspapers and drinks containers – had a solution in place. It could be recycled or composted, even if rates of actual recycling and composting were low. As recycling and composting ramped up, the diapers looked on track to become a larger and larger proportion of what was being buried and burned.

Scott Stewart, still in his windowless office, found his phone was ringing off the hook. 'We were being overwhelmed at the time by enquiries from the media in every part of the country,' he says. 'The company knew it needed to stand up, and strategically it wanted to be seen as part of the solution.'

*

On 20th June, 1989, P&G held a press conference at the National Press Club, a short walk away from the White House in downtown DC. The storied institution had counted every American president since William Howard Taft as a member and, over the years, its weekly lunches were attended by global political bigwigs like Winston Churchill.

'We had no idea who was going to show up,' remembers Stewart. 'Part of having it at the National Press Club was to establish the prominence P&G was assigning to the issue.' The night before the conference, Stewart, who had flown in from Cincinnati, tossed and turned in his air-conditioned room at the Marriott, unable to sleep. When, the next morning, he padded into the thickly carpeted room where P&G was holding the conference, it was packed with journalists and news cameras.

Richard Nicolosi – who led P&G's paper business, of which disposable diapers were a part – stepped up to the podium to address the room. P&G, he said, had decided to explore two highly promising potential solutions for diaper waste. First, the company would set up recycling projects in Washington, Minnesota, Wisconsin and Florida, washing and sanitising diapers before separating them into plastic and pulp. Second, said Nicolosi, P&G would run trials to see if it could compost its diapers.

The press conference created a stir. 'If P&G can find a way to routinely recycle disposables, it deserves a big hand from America's babies, their parents and Mother Earth,' wrote *Newsday*, whose reporters had taken a great interest in waste after closely covering the *Mobro* garbage barge saga.

Back in Cincinnati, Nicolosi and P&G's other executives created a battle plan to neutralise Lehrburger's study and eviscerate his claim that cloth diapers were better for the environment than disposables. P&G commissioned its own diaper study from Arthur D. Little. The Boston-based company, founded by a chemist, calls itself the world's first management consulting firm and is a long-time ally of the plastics industry. As evidence was mounting in the 1970s that the components of polyvinyl chloride caused cancer, the Society of the Plastics Industry hired Arthur D. Little to conduct a study. It showed that 2.2 million jobs depended on the PVC industry and claimed that a ban would amount to billions a year in losses. That study hit as journalists began writing about how chemical companies had known about the cancer-causing impacts of PVC for months but purposely chose not to disclose them. (PVC, incidentally, remains widely used today in everything from water pipes to credit cards, although a growing number of consumer goods companies are phasing it out.)

Arthur D. Little's diaper study found that cloth diapers consumed more energy and water, resulted in more air and water pollution emissions, and cost more than disposable diapers, even before cloth diaper service costs were included. Using Arthur D. Little's assumptions, it probably made more environmental sense for everyone's clothes to be disposable, deadpanned one critic.

It turned out that the consulting firm had taken some liberties. Tim Sergeant from Gerber, the cloth diaper maker, went to a conference at which an Arthur D. Little consultant presented the report's findings. Sergeant was baffled to discover that the firm had assigned a $6 labour cost to the act of throwing dirty diapers into a washing machine at home. He says it had also allocated all the machine's capital and maintenance costs to cotton diapers. Sergeant raised his hand and objected. He pointed out that parents could do other things while waiting for the washing machine to finish its load, and that washing machines were quite obviously used for clothes other than diapers. The presenter, he says, waved his misgivings away. 'He defended it very roughly,' says Sergeant, who still thinks about the episode 35 years later.

In August 1990, P&G distributed the Arthur D. Little study's findings via a letter to legislators. The letter went out under the banner of the American Paper Institute, a trade body. 'Dear Legislator', began a letter signed by API president Red Cavaney, 'The contribution of single-use diapers to the waste stream, and their consumer value, have often been topics of discussion. Unfortunately, many comments on this issue have been based on inaccurate or misleading information.'

Cavaney recounted how the study had found that single-use diapers offered better protection against diaper rash and decreased the spread of infection in day-care centres. There was no mention in the letter that the study being cited had been commissioned by P&G. In fact, there was no mention of P&G at all.

The following year, 1991, as a string of states considered requiring day-care centres to accept cloth diapers, another study broke. This one showed that cloth diapers could increase the risk of diarrhoea outbreaks in day-care centres. The study, conducted by researchers from the University of Texas Medical School in Houston, found an increase in diarrhoea-causing germs when several day-care centres used cloth diapers, though it didn't find an increase in actual diarrhoea outbreaks. Although P&G had contributed $80,000 to the research, there was, again, no mention of the company's sponsorship in the published study.

A flurry of environmental claims and counterclaims from the disposable and cloth sides followed, leaving already harried parents scratching their heads. Sorting through these claims – arrived at through complex industry-funded studies called 'life cycle analyses' – was wildly confusing. Key to the outcome of a life cycle analysis was the assumptions its authors chose. These included how many times cloth diapers could be reused, how often parents changed them, what temperature they were washed at, how much water was used to pre-soak the diapers before washing, and whether to include water used to flush poop emptied into the toilet from disposables.

Such life cycle analyses, or LCAs, were historically conducted to help companies make internal decisions on product design and manufacturing waste. The first-ever LCA was commissioned by Coca-Cola's head of packaging in 1969 when the company was considering using plastic bottles. But in the wake of the *Mobro*'s journey, LCAs were being commandeered by advertisers and manipulated based on the desired outcome. There was no standardised methodology and no accreditation, making it easy to cherry-pick the assumptions that were most favourable. Researchers largely relied on proprietary industry data and companies chose what they were willing to share.

When LCAs were used to compare cloth diapers against disposables, all they succeeded in doing was muddying the waters. Comparing different variables – like the manufacture of sodium silicate used for cloth diapers against the sodium hydroxide used for disposables – to crown an environmental winner was essentially impossible, and many researchers warned of the futility of drawing such conclusions. Unsurprisingly, the cloth and disposables camps also drew opposing conclusions over safety, comfort and price.

Decades later, as I write this book, little has changed. LCAs are widely used today by warring industries and are plagued by all the same problems. Parents with children in diapers, myself included, are as confused as ever.

*

In the early 1990s, armed with their new data, P&G's lobbyists fanned out across the country to quash the backlash against disposable diapers.

In New York, they tied up a brochure written by the State Consumer Protection Board that explained the pros and cons of cloth and disposable diapers. When the brochure was finally printed more than two years later, it was a shadow of its original self. 'The governor's office didn't want to piss off P&G and they kept making it weaker and weaker,' says Pat Rodriguez, back then the protection board's 32-year-old general counsel.

In Kentucky, Roger Noe, a community college professor and state legislator, also came up against P&G when he proposed levying a quarter of a cent tax per diaper to fund environmental cleanups. Every time the Cumberland River flooded, Noe found the trees in his neighbourhood, Harlan, covered with disposable diapers. 'We called them diaper trees,' he remembers. 'There were thousands and thousands of diapers.' But Noe's colleagues baulked at the idea of a diaper tax. 'They kind of made jokes about it, especially the conservative folks. They'd say, "You're full of it like a diaper is."' The biggest pushback came from disposable diaper makers. 'The P&G people were kind of stirred up,' says Noe mildly.

Lawmakers also proposed diaper taxes in Colorado, Connecticut, Wisconsin and Illinois, but P&G's lobbyists derailed all of these, arguing that they unfairly singled out diapers and penalised lower-income consumers. Proposals in California and New York for warning labels carrying messages about the environmental impact of disposables suffered the same fate. Testifying at a New York hearing, Carl Lehrburger found himself up against half a dozen P&G scientists, all armed with PhDs. 'It became clear that P&G was going to spend any amount of money to counter any legislation impacting diapers,' he says.

Kimberly-Clark's lobbyists and marketers also worked the phones, mobilising 100,000 parents in Wisconsin to defeat the state's proposed diaper tax. The company sent pro-disposables flyers to shareholders, and millions of leaflets to households across the US. 'We obviously spent a lot of time on this,' says Wayne Sanders, who was heading Kimberly-Clark's diaper business at the time. 'Were a lot of things filling up landfills? Yes. Were diapers a new category that didn't fill up landfills before? Yes. Could they be a poster child for filling up landfills? Yes. You had to deal with that.'

But it was P&G that took the lead. Between Pampers and Luvs, P&G had far more wrapped up in babies' bottoms than its rival. P&G was also the first consumer goods company to fully grasp the seriousness of the public's antipathy to plastic. In March 1991, P&G's president, John Pepper, stood shoulder-to-shoulder with DuPont's chairman to announce a plastics industry recycling goal, saying he had seen 'a real sea change' in Americans' environmental concerns. 'I have never seen an issue change with such force,' Pepper told the *New York Times*.

P&G was also the first in its industry to become a member of the Council for Solid Waste Solutions, which, two years after its inception, had grown to include more than 20 members making or using plastic. 'I think it was the ubiquity of their products,' says Don Shea of P&G's involvement. 'They wanted to play offence, not defence.'

By the late 1980s, thousands of public schools across the US were facing major budget crunches. P&G spied an opportunity in the harried teachers struggling to come up with lesson plans. Just like McDonald's, it began writing and distributing teaching kits for free. In 1989 alone, it spent $3.75 million on 'educational' programmes for use in high schools.

In one teaching kit called *Decision: Earth*, P&G asked teachers to get their seventh to ninth graders to compare the environmental impacts of cloth diapers against disposables using a 'fact sheet'. Students were given data showing that on four out of five parameters – energy, water pollution, air pollution, water consumption and solid waste – disposable diapers scored better than cloth. P&G didn't disclose that the data came from a study sponsored by the Diaper Manufacturers Group, the trade body for the disposable diaper industry. It also omitted to mention that it, P&G, made diapers.

After students had totted up how much worse the environmental impacts of cloth were, teachers should then stimulate class discussion, suggested the P&G teaching kit. 'Ask students to evaluate qualitative considerations,' it instructed teachers. 'What is the value of time to a working parent? What are some of the additional costs involved with the use of cloth diapers: pins, plastic pants, additional thicknesses needed as the child gets older, detergents, softeners, soaking liquids, diaper pails and disinfectants.'

P&G was essentially indoctrinating the next generation of disposable diaper buyers. The lesson plan provoked an outcry from environmentalists, who said it was biased and blatantly promotional to P&G. Even members of the advisory panel that P&G had used to vet its claims complained, with one saying he had 'red-marked' the diaper section but the company had ignored his suggestions.

But, by then, the wheels were in motion. Over the next two years, *Decision: Earth* was distributed to 70,000 teachers, reaching millions of schoolchildren across America.

*

In May 1990, 800 Seattle-based children under the age of three were picked to become guinea pigs in a P&G-led diaper recycling experiment.

Their parents were given special plastic-lined garbage cans for used diapers. Each week, trucks trundled up, collected the diapers and took them to a warehouse. There, the unopened bags whizzed up conveyor belts, dropping into a pulping tank containing water and a sterilising chlorine compound. The machine acted like an enormous blender, shredding both the plastic bags and the diapers. The shredded material was rinsed and the faeces diverted into a sewer. The paper fibres were sucked out, washed to remove the highly absorbent 'super slurper' crystals and then separated from the plastic using a vacuum process.

P&G said the plastic could be turned into flower pots and garbage bags, while the padding made from wood pulp could be used for cardboard boxes and building insulation.

Journalists, invited to tour the facility, were impressed by what they saw. 'Relying on the seats of their collective disposable pants, 1,000 Seattle babies may soon write a new chapter in recycling history,' wrote a reporter for the *Seattle Times* at the start of the experiment in February 1990.

But – just as with McDonald's clamshell recycling trials – the diaper recycling effort quickly ran into problems. While disposable diapers could technically be recycled, the economics of doing so in real life didn't stack up, particularly for the plastic components. The recovered plastics, although they could be used to make new products, commanded just 2 cents a pound. That just about covered the cost of shipping them to buyers. It was nowhere near enough to cover the cost of recycling.

There were other problems too. Superabsorbers lingering in the recycled pulp could cause flaws in the paper being produced, while potential pulp buyers were reluctant to make new products from old diapers. 'It appears that the costs of operating the facility for a single product such as disposable diapers will prove uneconomical,' a P&G executive admitted to the *Seattle Times* in January 1991.

*

Norma McDonald was sitting in the audience at P&G's annual general meeting on 9th October, 1990 when CEO Ed Artzt told the gathered investors and employees that P&G would make a compostable diaper as soon as possible and spend $20 million to promote composting across the world.

McDonald, who worked in procurement for P&G, was taken aback. 'At that time, boy, there's no composting industry per se and certainly not a composting industry that would take something that looked like a compostable diaper,' she says. 'But he was known for setting out extremely challenging objectives.'

McDonald had spent the better part of the past year looking for compostable alternatives to Pampers' plastic backsheet – the outermost part of the diaper that's used to keep moisture in. She sent the prototypes she found to a composting facility in St Cloud, Minnesota to see if they'd break down. Most didn't. The ones that did had other problems. One promising compostable film flopped when tested on babies in real life. 'These little kids were moving around in their sleep and the film was too noisy, it woke them up!' says McDonald. Another was too smelly. Compostable plastic, it turns out, readily allows odours through.

It wasn't just the backsheet McDonald was preoccupied with. The topsheet, tabs and superabsorbent polymers all needed to be replaced with versions that would quickly and fully break down in composting facilities. 'There are over 1,000 variations of polyethylene compounds and there were only three primary biodegradable polymers to choose from,' says McDonald. 'Not enough to replicate all the required properties.'

Still, the cellulose – which made up around 80 per cent of the diaper – did break down and so, in the spring of 1990, P&G began testing diaper-infused compost. It trucked this to Christmas tree farms and to the Sand Plain research farm in Becker, Minnesota to grow potatoes. It took it to a landscaping yard in the city of Elk River to grow grass, and to an iron range in northern Minnesota where the state's government was trying to reclaim mined land.

Overwhelmingly, the results showed that the compost retained far more moisture than a control compost with no diapers in it. It turned out that the superabsorbers that made the diapers so good at preventing leaks hadn't degraded. 'These are carbon-carbon backbone polymers that are fundamentally not biodegradable,' explains Chuck Pettigrew, a senior P&G scientist who worked on diaper degradation among other projects at the company for over three decades. 'Similar to microplastics, they are "forever" chemicals that will end up persisting in our ecosystems until the end of time.'

*

It's crazy to think that a multinational diaper maker had a big hand in the founding of America's modern industrial composting industry – but it's true. So determined was P&G to find a way to compost its diapers that, in 1990, it funded and organised the country's first trade group for composting companies: the Solid Waste Composting Council. A P&G executive even wrote the composting industry's first training manual.

Back then, America's composting industry was virtually non-existent. 'In the early '90s, when you heard the word composting in the US, you thought of a farmer with a manure pile or of people who had victory gardens for World War II and had little backyard composting operations,' says Carla Castanegro, a composter who was one of the Council's early members. 'There was truly nothing in between those.'

A small number of commercial composting facilities had sprouted across the US since the mid-1950s. But, by 1972, only one remained, down from a peak of 18. As with recycling, cost posed the major hurdle. Compost couldn't compete on price against more effective chemical fertiliser. Despite this, P&G saw composting as a serious substitute for landfilling. Composting, Artzt told P&G's shareholders, could turn up to 60 per cent of all residential waste into nutrient-rich humus. To make it economically viable, he was advocating that newspapers, magazines, telephone books, paper and cardboard packaging all be composted.

To save on collection costs, introduce economies of scale and ensure its diapers weren't excluded, P&G advocated mixed-waste composting. This required facilities to accept all kinds of trash, not just food and garden waste, belatedly sorting out what could be composted.

Jim McNelly, who ran the St Cloud composting facility, remembers pulling out mattresses and tyres from among the residential trash that came in. 'We called them the "goofies", because they were the stuff that would goof up the system if they got past,' he chuckles. Not everything was easy to sort. Small batteries – which McNelly calls 'little toxic drums' – often slipped through the sorting process, meaning dangerous heavy metals were found in the resulting compost.

A divide soon emerged in the composting community. P&G was convinced mixed-waste composting was the best way to go, since consumers had the least work to do and wouldn't have to separate out smelly diapers. But many composters and environmentalists worried that toxic heavy metals, shards of plastic, glass and grit could make their way from household trash into the food chain via compost used on farmland. Instead, they advocated for source-separated composting – where consumers separately bagged food and yard waste.

Another point of difference was that while some consumer goods companies began pushing composting as the best way to deal with their used cardboard and paper, environmentalists firmly believed that recycling paper was a much better use of resources, since it created the raw material to make new products.

Even as P&G was battling to shape the US's burgeoning composting infrastructure to its advantage, and simultaneously working to make a fully compostable diaper, another problem was emerging: how to handle the poop? Composting technology designed to process yard waste wasn't equipped to deal with this. 'At first, everyone concentrated on the diaper, not what was *in* the diaper,' says Castanegro. 'But even if a diaper could be made fully compostable, states were not going to allow something with poop and pee to go to a traditional composter.'

*

In the spring of 1991, even as P&G's own researchers were starting to seriously doubt the viability of ever making a fully compostable diaper, the company began running advertisements touting its composting efforts. The magazine ads, funded by P&G, showed photos of soil and saplings with headlines saying, 'Ninety days ago, this was a disposable diaper' and 'This baby is growing up in disposable diapers'.

P&G began running TV ads in Chicago, Denver and Sacramento, saying its diapers could be composted. In the state of Rhode Island, P&G sent households free samples of Luvs with messages saying: 'This product is compostable in municipal composting units. Support recycling and composting in your community.' Yet, at the time, just a small handful of US cities had municipal composting facilities. Chicago, Denver, Sacramento and Rhode Island had none.

P&G's ads caught the attention of a taskforce of ten attorneys general, which launched an investigation into the company's claims. By November 1991, the taskforce had concluded that P&G was misleading American consumers. 'By promoting their disposable diapers as compostable when facilities that accept diapers for composting are virtually unavailable, Procter & Gamble is deceiving consumers,' said New York Attorney General Robert Abrams in announcing a 'green collar fraud settlement' with the company. Abrams pointed out that building a single composting facility would cost millions of dollars. 'While we encourage companies that wish to lobby for the development of composting to do so, it is misleading to market products as if the solution were already here.'

P&G agreed to modify its composting ads. But, as with its teaching kit, by then millions of Americans had seen the ads. Many had read newspaper articles carrying the findings of the pro-disposables studies P&G had funded. P&G had also handed out 14 million copies of a six-page brochure in colour called *Diapers and the Environment*, in which it claimed that its composting trials at the St Cloud facility were having 'very positive results'. It brazenly attached discount coupons for Pampers and Luvs to the brochure.

Parents, once conflicted, now felt reassured that disposables, if not any better than cloth, at least weren't any worse.

'If I were to describe who the target was, it was the guilt-ridden mother of small children,' says Charlie Cannon, who served as the Solid Waste Composting Council's first head. 'The purpose of all this research, marketing and development was to make people feel good about disposable products, whether in the food industry or personal care or anything else.'

By now, few news outlets were still preoccupied with diapers. In August 1990, Iraq had invaded Kuwait. A few months later, in early 1991, the US and its allies unleashed Operation Desert Storm, a major air campaign targeting Iraq. 'With the situation in Iraq and Kuwait, people moved on just like they move on from everything else,' says Scott Stewart.

The proposed state bans and restrictions on disposable diapers fizzled out. By 1992, cloth diaper sales were declining again; Gerber, the cloth diaper maker, said it would close three weaving plants and lay off 900 workers. 'P&G's Arthur D. Little study assuaged consumers' concerns about using a disposable product,' says Gerber's Tim Sergeant. 'It provided some justification for the consumer to say, "It doesn't matter since cloth diapers also have some environmental impact." We talked about it as relieving the consumer's conscience.'

P&G's $20 million investment in composting meant the company continued researching compostable diapers for years after. But as the backlash against diapers ebbed, P&G declined to invest further, leaving the composting council to find its own way. It eventually shelved the idea of making a compostable diaper. 'It's one thing to say you're using a biodegradable material, it's another to design it and build it so it actually works,' says Chuck Pettigrew, the former P&G scientist. 'At the end of the day, the whole idea of creating a biodegradable diaper we realised was going to be too expensive.'

By 1996, a whopping 94 per cent of American parents relied solely on disposable diapers, an all-time high. New York City, home to eight million people, didn't have a single diaper service company left within its five boroughs. The membership of the National Association of Diaper Services had halved from five years earlier to stand at 90 members nationwide. Today the association no longer exists.

The decline of the diaper washing companies made choosing cloth diapers far more inconvenient and hastened their demise. It also eroded a key environmental benefit: lower carbon emissions from washing and drying diapers centrally using machines that are more efficient than ones for home use.

There's more to this than disposables pushing cloth diaper service makers into extinction. The disposable diaper fundamentally changed children's behaviour so that they stayed in diapers far longer. The superabsorbers meant children didn't feel wet in disposables the way they did in cloth, eroding one of the strongest incentives for them to embrace potty training.

In 1957 – before Pampers burst onto the scene – 92 per cent of American children were toilet-trained by the age of 18 months. By 1999, that ratio had essentially flipped and 96 per cent of American children were *not* toilet trained by the age of two. To emphasise, that means just 4 per cent of two-year-olds were toilet-trained.

Of course, the fact that more women were working rather than staying home all day had an impact. But it seems clear that at least some of this comes down to the millions P&G poured into convincing parents first to abandon cloth and then of the dangers of 'rushing' children into potty training. It also comes down to the disposables themselves. They were effective, ultra-convenient and cheap. Once in the trash, neither parent nor child gave them a second thought.

PART II

THE FIGHTBACK

CHAPTER 5

HEFTY'S CLAIMS TURN TO TRASH

'It's that public relations value that has to be considered as opposed to real solutions'

The sun had already set over Washington, DC when the fax machine at Collier, Shannon & Scott lit up and began spitting out a noisy stream of pages.

The sound stirred John B. Williams, who had been reclining precariously in a swivel chair after a long day drawing up strategies to defend Mobil, the chemical giant who had recently signed on as a client. Williams scrambled to his feet to rifle through the first arriving pages – a lawsuit filed by the pugnacious attorney general of Texas, Jim Mattox. It demanded that Mobil stop marketing its Hefty trash bags as 'degradable' and that the company pay penalties.

Mobil had promised that by buying Hefty trash bags, Texans would be helping the environment. This was a deceptive claim that amounted to a cynical manipulation of people's environmental fears, said the complaint. 'All Mobil truly sold them were false hopes,' it added.

The date was 12th June, 1990. Mobil had foreseen legal action, several weeks earlier hiring Collier, Shannon & Scott. Thirty-eight-year-old Williams, a seasoned litigator who had previously represented Mobil's parent company in a price-fixing case, had stayed late in the office that evening, anticipating the delivery of a formal complaint. What surprised Williams was that the pages just kept coming and coming. 'My reaction was, and I think we joked about it, "Maybe we should unplug the fax machine",' he says.

One after another new complaints were spat out until Williams was holding lawsuits from attorneys general in six other states: Minnesota, New York, California, Wisconsin, Washington and Pennsylvania. They all accused Mobil of deceptive advertising.

*

As the saying goes, in the midst of every crisis lies great opportunity. For corporate America, the country's perceived landfill crisis was no exception.

Three years after the *Mobro*'s load was burned, a growing number of companies had pivoted from hunkering down in damage control mode to launching new products and advertising to capitalise on the public's desire to be green. Much of what was being said and sold was misleading. The green claims drew the attention of the state attorneys general who decided – much like Will Collette had in targeting McDonald's – that making an example of market leaders like Mobil and P&G was the best way to get other companies in line.

The chemicals arm of one of the largest oil companies in the world, Mobil had begun printing the words 'NEW DEGRADABLE' on boxes of Hefty trash bags in June 1989. The large trash can liners were sold at major supermarkets across the US. Overnight, the boxes also began featuring a bald eagle flying across the sun. The eagle made another appearance on the back alongside the slogan 'Hefty

Helps!'. The new box also featured a pine tree and an explanation that the bags would degrade without harming the environment.

Mobil claimed that unlike compostable plastic – the kind P&G was trying to use – its degradable plastic didn't even need to be sent to composting plants to return to Mother Earth. 'Once the elements have triggered the process, these bags will continue to break down into harmless particles even after they are buried in a landfill,' promised Mobil. What made these claims so notable was that the company knew, and had loudly said, that degradable trash bags were a total farce.

For months before launching the bags, Mobil had gone to great lengths to point out that plastic bags and other allegedly degradable plastic products going to landfills weren't actually breaking down and were not a solution to America's solid-waste problem.

But consumers loved the idea of degradable products. Degradability was the ultimate antidote to the evils of consumerism, allowing people to shop 'til they dropped, guilt-free. Legislators loved the idea of degradability so much that several wrote bills requiring companies to ensure the plastic rings that held six-packs of drinks cans together were degradable.

'Degradable was a feel-good word,' says Mike Levy, who served as Mobil's chief lobbyist at the time. 'For some people, degradable means it magically disappears and there's this idea that if something is made with degradable materials, it's not a problem.'

But, of course, it was a problem.

To explain why, I need to step out of the Mobil story for a minute to explain a couple of things.

*

Companies in the US were selling two main types of degradable plastic products back in the 1980s and early '90s. The first is photodegradables – plastics designed to break down when exposed to sunlight for a sustained period. The second is biodegradables – plastics that decay naturally into carbon dioxide, water and biomass as micro-organisms consume their carbon. They're often made from bio-based materials[1] like sugarcane or starch, but a small slice of fossil-fuel plastics can also biodegrade.

Today, consumer goods brands rarely make claims about photodegradable plastics because it's impractical to design packaging intended to be disposed of in full sunlight. Photodegradables may also break into smaller fragments – microplastics – instead of fully degrading. In Europe, for a time some companies tried selling 'oxo degradable' plastics, which combine conventional plastics with an additive that promotes degradation into smaller pieces. However, such plastics don't fully degrade and the European Union has moved to ban oxo degradables. I'll tell you more about microplastics in Chapter 12, but for now it's enough to know that back in the early 1990s, few people were talking about the problems posed by tiny plastic particles.

'Biodegradable' plastic continues to be a widely used marketing term today. It's also widely misunderstood and misused. Labelling a plastic biodegradable says nothing about *how long* it takes to fully break down – a littered biodegradable plastic bag that remains intact for five years can still entangle wildlife and leach microplastics and chemicals into the environment over this period. For this reason, a handful of US states have banned the use of the word 'biodegradable' in marketing language.

It's difficult to guarantee that a biodegradable product will break down in landfills, because factors influencing degradation, like

---

[1] About 44 per cent of bioplastics (measured as a share of global production capacity of bioplastics) are not biodegradable, according to European Bioplastics, a trade body.

moisture content and temperature, vary widely. Many environmental engineers say it's undesirable to put degradables into landfills since they can release methane, a highly potent greenhouse gas. Although Lowell Harrelson saw methane as a potential goldmine since it can be turned into energy, methane capture is expensive and inefficient. Today, only some of the methane coming out of landfills is captured.

Then there are compostable plastics. Compostables – like the diapers P&G wanted to make – are designed to biodegrade in a *specific environment*, such as an industrial composting plant that uses high temperatures, and within a *specific time period*.

Take a deep breath and read this twice: *all compostable products are biodegradable in a specific environment but not all biodegradable products are compostable.* Compostables are essentially a much more defined and demanding sub-category of biodegradables.

Today, compostable products and packaging are mostly landfilled or burned because many consumers don't have easy access to industrial composting facilities that accept these. Putting compostables in general waste bins where they'll go to landfills is a bad idea. They'll remain intact for decades and not save any landfill space. Or if they do break down, they'll release methane.

I've included more information on bioplastics, biodegradables and compostables in the FAQs at the end of the book. I urge you to read it. Biodegradable and compostable plastics are the most misunderstood area I've come across – and the most greenwashed.

*

Okay, so now that you have the basics, back to Mobil.

Although Mobil had for years conducted research at its labs in Edison, New Jersey into degradables as an alternative to conventional plastic, the company – along with many of the plastics industry's biggest players – had been staunchly against commercialising these.

For one thing, scientists harboured fears that the new plastic could decay before it was meant to, transfer smells to food or infect it with bacteria. There was also the concern that chemicals from decomposing plastic could leach into groundwater, or that the methane it released could create fires in landfills. Thrown in recycling bins, degradable plastic could contaminate plastics recycling processes, having a knock-on effect that was actually worse for the environment. Then there was the small matter of expense. The industry had poured billions of dollars into plants and equipment to make conventional plastic. New equipment required, like silos and holding tanks, could add as much as 20 per cent to production costs.

The chemical companies' lobbyists also warned that, once consumers realised degradable plastic didn't generate more landfill space, there could be a backlash, triggering more bans and restrictions against plastic that couldn't be recycled.

The Council for Solid Waste Solutions had publicly denounced degradability, saying it wasn't a solution for America's solid-waste problem. 'Modern landfills are designed to retard or limit the decomposition process of degradable materials,'[2] it explained in a 1989 book. 'Plastic products in the waste stream resist degradation and do not easily decompose.'

Even as it explored compostable diapers, P&G was so convinced that diapers marketed as biodegradable weren't a solution to waste that it poured $160,000 into discrediting brands that claimed their diapers were biodegradable. It paid for a full-page ad in the *Boston Globe* pointing out that biodegradable diapers don't actually break down in landfills. 'There is no such thing as a biodegradable diaper,' said the ad.

---

[2] This is to reduce the release of methane.

Mobil too went on the offensive, organising five symposiums across the country to tell anyone who would listen that degradable plastic would do nothing to help the US with its solid-waste problem. Mike Levy, the company's chief lobbyist, pointed out that the additives used in degradable bags weakened them so companies needed to use more plastic to make them as strong as regular bags. Companies selling degradable bags were charging consumers more for an allegedly eco-friendly option. The irony was that the bags could take up *more* space in landfills. 'When the consumers find out that this extra cost isn't going to do what they expect . . . we don't want to be behind that,' Levy told a reporter in July 1989.

At one of the symposiums, a top official from Mobil – who also represented it on the Council for Solid Waste Solutions – laid out the company's stance. 'Mobil has concluded that biodegradable plastics will not help solve the solid-waste problem,' said Robert Barrett. 'We do, however, see that there are some short-term public relations gains in switching to a photodegradable plastic grocery sack or consumer trash bag, or even a biodegradable bag of each type,' he continued. 'And it's that public relations value that has to be considered as opposed to real solutions to the problem.'

*

The now-ubiquitous large polyethylene trash bag was dreamed up by a Canadian inventor in 1950.

It was intended only for commercial use; the first batch was sold to Winnipeg General Hospital. But by the 1960s, as the amount people threw away soared across North America, Union Carbide began marketing the bags to housewives under the Glad brand. Much like how the day-care centres had abandoned cloth diapers, making parents put their babies in disposables, US sanitation departments had begun pressing residents to use lighter polyethylene trash bags

115

rather than the metal garbage cans their workers otherwise had to empty into trucks.

Glad embarked on a bout of 1970s advertising that embodied America's changing attitude to garbage. 'Pack up your troubles in a new Glad bag and smile, smile, smile,' went the catchy tune for an ad showing a gleeful family ridding itself of bulky bags of trash.

While consumers previously dragged the heavy metal cans to the kerb or gingerly carried out open paper bags, the lightweight, durable plastic bags made it a breeze to throw things away and never think about them again. Plastic trash bags to hold America's – increasingly plastic – discards exemplified a culture of consumerism that conflated excess with success and a casual carelessness with being carefree. The bags allowed people to truly disconnect from their discards – out of sight was out of mind. Between 1960 and 1990, the total amount thrown away by American households surged 136 per cent, the vast majority landfilled.

Hefty's own tagline was 'Tough enough to overstuff'. When it came to throwing things away, the implication was that the sky was the limit. Alongside the trash bags, Mobil sold a host of single-use plastics to fill them, like Hefty-branded cups and plates, and Baggies food storage bags.

In 1984, Hefty rolled out the Cinch-Sak trash bag. The yellow drawstring tape allowed people to fill the bag up to the top and easily tie it up. 'Fill it up and close it up. Reopen it and put in more!' enthused one ad for Cinch-Sak. 'Never touch garbage again!' said another.

Why stop at trash? Mobil encouraged consumers to use the disposable drawstring bags for laundry, beach supplies and carrying sports gear too. The disposable bags offered virtually limitless space. Consumers previously restricted to a single trash can could now throw away as many bags as they wanted. The company recruited Boy Scouts to sell its bags to fundraise for troop activities.

Through the 1980s, the number of trash bags households put out for collection kept climbing. Then, in 1988, as America's environmental concerns surged, degradable plastic bags from Mobil's competitors with names like LitterLess, Plastigone and Good Sense began popping up across the country.

From Mobil's perspective, one particularly irksome competitor was Poly-Tech, a Bloomington, Indiana company that made the Ruffies line of trash bags. Ruffies had been on a tear ever since Poly-Tech broke with industry tradition and abandoned the cardboard boxes that trash bags across America had long been sold in. Ruffies began showing up on shelves in rolls packed inside colourful plastic sleeves. Axing the box meant Poly-Tech could price the bags 15 per cent lower than rival brands while also offering retailers a fatter profit. Ruffies' sales jumped from fewer than half a million dollars in 1977 to more than $12 million by 1983.

Ruffies had received another huge boost when Poly-Tech signed Jonathan Winters to endorse the brand. Winters, whose portly face brought to mind a sad Basset Hound, was one of America's best-known comedians and had for years done TV spots for Hefty trash bags. He wore a pristine white uniform and played a well-heeled garbage man sporting a British accent whose job it was to collect 'garbahge'. But Mobil fired Winters in 1977 for reasons it never revealed, and now Winters repped for Ruffies and poked fun at Hefty. 'Because Ruffies cost less, you could save fifty thousand dollars,' he said in one ad, while shaking a broken calculator. 'Ruffies: Big tough bag, little sissy price.'

With Winters' help, Poly-Tech was selling 1.26 billion bags each year. It then announced it would make photodegradable Ruffies. Poly-Tech promised they would take up less space in landfills and break down in oceans and on roadsides if littered. The bags included an additive that triggered the degrading process once the plastic was exposed to sunlight for several days. What Poly-Tech didn't say was that there was no sunlight in landfills, while any littered plastic trash bags that did break down elsewhere would release billions of microplastics.

When reporters phoned Mobil to ask if they'd be following suit, the company's officials attacked Ruffies' move. Barrett said the company wasn't convinced that a degradable plastic bag was 'truly good for society' or a solution to America's waste problems. He also underscored that for the bags to degrade, they'd need to be left above ground.

Mobil spokesman Allen Gray went one step further in denouncing Ruffies. 'The degrading process takes far, far too long to create extra space in landfills,' he said. 'The money is much better spent on real solutions to the problems.'

*

Despite Mobil's seeming disdain for greenwashing, by early 1989, Hefty's marketers were increasingly convinced that the brand needed to launch its own degradable bags.

It was clear that consumers had a big appetite for technology that would allow them to keep using throwaway plastic, guilt-free. Ruffies, shortly after the launch of its degradable bags, announced that demand was 'overwhelming'. A small diaper maker that switched to degradable plastic reported sales had grown 900 per cent in just four months.

At first, Mobil's marketers floated the idea of a new degradable variant of Hefty that left the original untouched. But as consumers became increasingly irate over plastic waste, Mobil made an astonishing about-face. It decided to switch wholesale to degradables, reformulating the existing bags to include an additive that helped them degrade in sunlight. 'At the end of the day, there was a lot of pressure on Hefty bags,' says Mike Levy, the Mobil lobbyist. In the absence of strong legislation outlining what companies could and couldn't say, 'this whole area of marketing and environmental claims was wide open'.

Switching to degradable plastic would also protect Mobil's lucrative plastic grocery bag business, home to brands like Kordite, Baggies and Marketote.

Plastic grocery bags, which had burst onto the scene in the late 1970s, counted as one of the industry's biggest success stories. In 1982, just 5 per cent of bags bought by American grocery stores were plastic. By 1990, plastic made up a whopping 60 per cent of these bags. But just as the cloth and disposable diaper makers had jousted over the past few years, paper and plastic bag makers had engaged in an escalating tit-for-tat, trading claims and counterclaims about whose product was better, cheaper and kinder to the environment.

Plastic bags represent progress, said plastics makers. Paper bags are as American as apple pie, said the paper camp. Plastic bags are cheaper, said the plastics industry. It may seem that way, said paper makers. But plastic bags take 18 per cent longer to pack and packers often double-bag, which makes plastic bags 10 per cent more expensive in practice. Paper bags are hard to carry, said plastics makers. Heavy plastic bags hurt people's hands, shot back paper makers. You get the point.

One claim the paper industry smugly and loudly made that plastics makers hadn't been able to counter was that paper bags were degradable. Until now.

Mobil's surveys had shown that consumers didn't fully understand degradable plastic. The company's marketers decided to capitalise on their confusion. 'You're getting good answers to support going down that road because of a lack of understanding on a scientific basis about degradable products and the environment,' says Levy who, with the benefit of hindsight, is ultra-candid about Mobil's intentions. 'It was playing on the public's opinion, the average public understanding.'

Levy watched from the sidelines as Mobil began marketing its trash bags as degradable. He says that the decision was taken against his advice, and even back then he didn't hold back on expressing his ambivalence. 'We're talking out of both sides of our mouth,' he told a journalist soon after Mobil launched the NEW DEGRADABLE Hefty box. 'Degradability is just a marketing tool,' he added. 'I don't think that the average consumer knows what degradability means. Customers don't care if it actually solves the solid-waste problem. It just makes them feel good.'

*

The barrage of lawsuits simultaneously filed against Mobil had been coordinated by an assistant attorney general in Minnesota named Doug Blanke. Blanke had dark hair that he liked to offset with luridly colourful ties. He had harboured an activist streak since his days as an undergraduate at Michigan State in the early 1970s. Inspired by Ralph Nader's call for students to fund their own public interest research groups – PIRGs – to fight for social causes, Blanke helped found the university's first such group.

He graduated top of his class, went on to Yale Law School and eventually landed at the office of the state's attorney general, Hubert H. Humphrey III, where he headed up the consumer protection unit. Looking around in the late 1980s, Blanke saw a free-for-all as marketers rushed to slap unfounded green claims on their products. 'If a term gained popular currency, then companies would claim their products had it,' he remembers.

Blanke went to Humphrey, whom everyone called Skip, making the pitch that cracking down on false green claims should become a north star for their department. His boss needed little convincing. Humphrey's wife had only recently brought home a box of Mobil's degradable trash bags from the supermarket and the attorney general, along with everyone else in his household, was briefly taken in by the claims.

'We were proud until we began thinking about just where those trash bags would really end up, and about the fact that when we put them in the trash, they headed off to landfills and incinerators where they would never have a chance to degrade,' Humphrey says. Being bamboozled by Hefty brought home to the attorney general just how out of control things had become.

A staunch liberal and the son of former US Vice President Hubert H. Humphrey, the younger Humphrey saw himself as Minnesota's antithesis to Ronald Reagan, fighting on behalf of the little guy. Since becoming attorney general in 1983, he had investigated everything from child sexual abuse to the rigging of highway construction bids. He saw taking on Mobil as a chance to put corporations engaging in greenwashing on high alert. 'The companies were making all these claims,' he says. 'We wanted to make sure their claims were honestly stated and backed up with facts and information.'

Humphrey incongruously parked a large shopping cart in his office in Minnesota's white marble capitol building in St Paul. He got staffers to fill it with products dubiously marketed as green. There were plenty to choose from. Between 1989 and 1991, the percentage of new products popping up in grocery stores making environmental claims more than doubled to stand above 12 per cent. A September 1990 survey of US companies by Deloitte and Stanford's business school found that 31 per cent had conducted special promotions or run advertisements with an environmental theme.

When visitors enquired about the cart, Humphrey would pull out the most brazen offenders, launching into a tirade against the misleading claims being made. 'It's gone too far,' he'd say. Polystyrene fast-food containers were labelled as recyclable, even though no Minnesota facilities would take them. Hairspray makers – who several years earlier had been banned from using ozone-depleting chlorofluorocarbons – were suddenly marketing the sprays as 'ozone-friendly'. Dozens of companies were labelling plastic products intended for landfills as degradable.

Blanke prepared for his assault on misleading degradables by phoning assistant attorneys general in Texas, New York, California and three other states. He asked if they'd be up for investigating the sellers of degradable plastic, particularly Mobil Chemical whose plastic bags had helped its parent Mobil Oil rake in $56.7 billion in sales the previous year.

Several of the other attorneys Blanke called had worked with him before. Attorneys general across the US had begun banding together in the 1980s after Ronald Reagan kicked off a deregulation crusade that effectively yanked big government off the backs of companies. Reagan's policies inadvertently created a regulatory vacuum that allowed companies to make an array of unfounded marketing claims.

The man known as Reagan's 'deregulation czar', the conservative economist James C. Miller III, had served as head of the Federal Trade Commission through the first half of the 1980s. The FTC was intended to be the defender of consumer rights, but Miller watered that focus right down. In the decade to 1987, its resources for advertising enforcement declined 42 per cent and there was a mass exodus of its top talent to private law firms or other public agencies.

'The FTC during the '80s was moribund. Someone I worked with back then said their name should be "forget the consumer",' says Steve Gardner who, as assistant attorney general for Texas, led the state's case against Mobil on behalf of attorney general Jim Mattox in 1990. 'They were understaffed, underfunded and under Reagan. We were pretty much fed up with them.'

Mattox famously backed Gardner even when he sued Quaker Oats, one of America's most-loved companies. Quaker's ads claimed that eating its oatmeal could reduce cholesterol and the risk of heart disease. Taking on Quaker proved hugely unpopular and some said the foray cost Mattox his bid to become Texas governor. But Quaker eventually changed its ads and agreed to donate $75,000-worth of food to food banks in penance. 'Jim let me sue anybody,' says Gardner. 'It was great.'

Back in the 1980s, Gardner's office on Jackson Street in downtown Dallas, a stone's throw from the county courthouse, was filled with cereal boxes, baby formula, auto parts guides and other memorabilia from past cases. The son of a Dallas police officer, Gardner openly lambasted his opponents with none of the usual restraint shown by his colleagues. 'It's not fat-free, it's truth-free,' he told a reporter about claims made by Kraft about its mayonnaise.

Mobil, Gardner felt, had no qualms about misleading people. 'One huge problem with Mobil's damn bags was consumers thought they could put 'em in their garbage and they would degrade,' he says. 'It was a bullshit, completely unnecessary and horrible claim.'

Blanke and Gardner found another strong ally in New York's assistant attorney general Andrea Levine. A short-haired brunette who sometimes clashed with Gardner over his outrageous comments, Levine had begun her legal career making $15,000 a year fighting on behalf of low-income New Yorkers. She could stomach the long hours and bad pay. But then she started becoming emotionally involved in cases involving child neglect and abuse. 'Once I had a kid, I just couldn't do it anymore,' she says.

A friend in the New York attorney general's office put in a good word and Levine was hired by the AG's consumer protection unit, where she was quickly drawn into working on advertising claims. The new job, she says, 'wasn't exactly curing cancer', but Levine felt a sense of satisfaction knowing that she was helping consumers in an area where they could be easily deceived. 'If I say my cereal is tasty and crunchy and you buy it and it isn't, you won't buy it again,' she explains. 'But on environmental claims, consumers can't judge the truthfulness for themselves.'

After getting wind of the AGs' investigation, Mobil stopped touting its bags as degradable. But the states decided to sue anyway. The company's switch to degradables after the many months it had spent discrediting them was seen as unforgivable by multiple attorneys general, who thought Mobil had brazenly profited from consumer concern. 'Their eyes were wide open,' says Levine. 'The irony was that at the same time that they were marketing degradable trash bags, they were sending out corporate missives saying degradability isn't the answer to the solid-waste problem.'

The AGs' ire was further stoked when Mobil refused to recall packs of bags that had already been distributed, leaving tens of thousands of boxes advertising the degradability of the bags on store shelves across the country, potentially for several more months.

It wasn't just the AGs investigating Mobil. The FTC was, too. After eight years of sitting on the sidelines, the agency had begun shaking off its stupor and was finally starting to pay attention to deceptive advertising cases under a new chair, Janet D. Steiger.

The widow of a popular Republican congressman who had died young, Wisconsin-born Steiger was, on the face of it, an odd choice. Neither a lawyer nor an economist – she had studied mediaeval literature – Steiger had a soft-spoken manner and a love of needlepoint. But she also had a strong record heading the US Postal Rate Commission, where she had gained a reputation for being pragmatic, even-handed and a fast learner. Another not-too-minor advantage was her personal link to George H.W. Bush – the president was the godfather of her son and had been a close friend of her husband.

Once at the FTC, Steiger quickly became interested in green claims, describing them in a 1991 speech as 'the hottest advertising issue around'. And so the FTC began its own investigation of Mobil.

As the complaints rolled in, John Williams recalls his client feeling unfairly singled out. 'Mobil felt quite put upon because they were not making the most aggressive claims, yet they were being sued,' he says. His argument was that his client's bags did include a chemical that would make them degrade, 'so we said "Yeah, it's true."'

Of course, as with companies' claims about compostability and recyclability, the key issue was whether the bag would actually do what the label claimed. In this case, it definitely wouldn't. It was clear that, in its eagerness to make money, Mobil had crossed a line. 'This was the start of the attack upon advertising that was true but misleading in terms of its effect,' says Williams.

*

Mobil eventually settled with the attorneys general and the FTC. But the most notable thing to come from the whole episode was a set of federal guidelines on how companies should make green claims to avoid running afoul of the US law that bans deceptive advertising. Today they're known as the FTC's 'Green Guides'.

When the attorneys general came together to investigate Mobil for its bags, P&G for its compostable diaper ads, and other companies, there were no federal laws or guidelines specifically addressing how companies should make green marketing claims. Section 5 of the FTC Act prohibited 'unfair or deceptive acts or practices in or affecting commerce', but the broad language – coupled with the FTC's 1980s malaise – meant that in practice many companies' green claims were intentionally or unintentionally misleading.

Even as they worked to build a case against Mobil, the attorneys general saw an opportunity to make a far bigger difference by forever changing the way companies made green marketing claims. Over two days in March 1990, Doug Blanke and AGs from other states lined up on a dais in a ballroom at the Radisson hotel in St Paul, Minnesota. Dozens of corporate executives, state officials and environmentalists came forward to offer their views on how environmental terms were being used, or misused, and how these might be regulated.

In November 1990, AGs from ten states published the 'Green Report', a set of recommendations for how companies could make green marketing claims without being deceptive or misleading. The report recommended that companies making environmental claims should be specific and avoid saying generic things like 'environmentally friendly'. Another recommendation was that products claiming to use recycled content should state the percentage of recycled material used, and also whether it was derived from waste generated by consumers or by factories, the latter being far easier and more appealing for cost-conscious companies to reduce.

Crucially, the 'Green Report' recommended that claims about a product being degradable or recyclable shouldn't be made unless the advertised disposal option was *currently available* to consumers in the area in which the advertisement was made. The idea was to move past claims that might be true on paper but in practice never materialised. 'It was a coordinated effort to move the market to a place where there were some guardrails about what was said when making environmental claims,' says Blanke.

The AGs saw their recommendations as a first step. They called for the FTC to formally adopt them as its own federal guidelines. They also wanted the Environmental Protection Agency to put in place rules defining key environmental claims, like 'recyclable' and 'degradable', that would have the force and effect of law. 'In all honesty, I'm not sure how much longer we can hold things together with the hit-or-miss approach of occasional lawsuits and settlements,' Humphrey told an environmental labelling conference in October 1991.

By then, three states had their own regulations governing environmental claims, and similar legislation was pending in at least six others. The states had all taken different approaches. On biodegradability, for instance, Pennsylvania was proposing to allow products to be marketed as biodegradable if they decomposed after any length of time, while in California, the term could only be used if the product broke down within one year. In Rhode Island, companies were prohibited from using 'biodegradable' in their product marketing at all.

For consumer goods companies like P&G, Kraft and Unilever, the growing patchwork of state-by-state regulation had sharply raised the risk of making any environmental claims whatsoever. They manufactured their products in a small number of locations for national distribution and often didn't know in advance where they would be sold. Modifying packaging claims to meet every state's regulation would be expensive and impractical.

Ross Love, P&G's advertising head, testified at FTC hearings that the emergence of state-by-state regulation could 'eliminate' all advertising and labelling about environmental issues. Another ad exec, Hal Shoup of the American Association of Advertising Agencies, testified that environmental marketing was fast becoming the third rail of advertising. 'Touch it and you die.'

While the consumer goods companies wanted national guidelines, they certainly didn't want ones written up by the state attorneys general. Bob Viney, P&G's first-ever global environmental marketing director, was part of an industry group that wrote up its *own* guidelines. 'Our pitch to industry was, "If you don't support our group's work to develop industry guidelines that you will voluntarily adhere to, you're going to find yourself regulated by government",' says Viney.

After months of work, on Valentine's Day of 1991, 11 industry trade groups filed their own, far less stringent, joint recommendations for how the FTC should regulate green marketing claims. The industry coalition – representing consumer goods, packaging, advertising and retailing firms – lobbied for guides rather than hard-and-fast rules, arguing that the environmental arena was dynamic.

Ironically, executives pointed to what they claimed was rapid progress in plastics recycling to bolster their case. 'I fully expect that, in the next few years, the recyclability of all common package forms will also be old news,' Tom Rattray, P&G's head of environmental quality, told an EPA conference in October 1991. 'Everyone will know it because everyone will be doing it.'

Under pressure from companies as well as the AGs, eventually, on 28th July, 1992, the FTC published the 'Green Guides'. Mary Engle, back then a young FTC attorney charged with drawing up the agency's guides, says they drew in part from the AGs' report and in part from the industry's suggestions.

Compared to the AGs' recommendations, the 'Green Guides' gave companies concessions on contentious issues, such as when a package could be labelled recyclable or compostable. The 'Green Guides' allowed companies to make a national claim – with no qualifiers required – that products were recyclable or compostable as long as collection programmes existed for 'a substantial majority' of communities. When consumers were subsequently polled, 46 per cent incorrectly thought the claim 'recyclable' meant collection programmes existed in their communities.

Where products could only be recycled or composted in a minority of communities, companies could still tout them as recyclable or compostable, as long as they included a qualifier asking consumers to check that recycling or composting facilities existed locally. In practice, this meant that even if less than 1 per cent of US municipalities accepted plastic bags for composting, manufacturers of such bags could nationally label the bags as compostable. Adding a qualifier – which many companies did in small print or on the back – only reduced the fraction of respondents who incorrectly concluded that collection programmes existed in their communities by three percentage points to 43 per cent.

The FTC's recommendations for companies making claims about degradability were similarly loose. Companies could make such claims if these were backed by scientific evidence to ensure their product would break down within a 'reasonably short time' when disposed of in their 'customary' location.

The industry hailed the 'Green Guides' as a win. 'The FTC took our guidelines and put them into regulations almost word for word,' Bob Viney says proudly. States incorporated the 'Green Guides' into their own laws or used them as the standard against which to determine liability and evaluate whether claims were deceptive.

With federal guidelines finally in place, and a seemingly resurgent FTC ready to hold companies to account, the state AGs began to pull back on policing companies' green claims, focusing their attention on other issues.

Skip Humphrey, Blanke and the others hadn't got everything they wanted, but at the time getting the FTC to put in place federal guidelines felt like cause for celebration. 'We weren't so naive as to think all the problems would be solved by federal action but, in Minnesota, we were thrilled,' remembers Blanke. 'We expected that the number of deceptive claims would reduce sharply.'

They didn't.

The new national certainty offered by the FTC's 'Green Guides' emboldened companies to keep making environmental claims and by 1995 the number of claims deemed deceptive by environmentalists had increased.

The FTC has updated the 'Green Guides' several times since then; however, many of the same issues that existed in 1992 have continued. Today, it's not unusual to be sold 'eco-friendly' bags, diapers and wipes that rely on the kind of general claims the commission warned against all those years ago.

Misleading claims about degradability in particular abound, as I referenced in the Preface. One example is the US dog poop bag industry, a major chunk of which consists of brands claiming to be biodegradable or compostable. They variously call themselves 'planet-friendly', 'eco-friendly' and 'landfill-friendly'. Most didn't return my emails, but one I spoke with acknowledged that composting facilities don't accept the bags and that they go to landfills. He claimed that they will break down there (it's very hard to guarantee this) and also that this is a good thing (it isn't, because of the methane).

Some US courts have looked to the 'Green Guides' when ruling in favour of marketers who say it isn't misleading to label their products as recyclable, even when they know these aren't being recycled in practice. As long as the products are *technically* capable of being recycled, courts have said the marketers didn't flout the guidelines.

Mary Engle – who worked for the FTC for nearly three decades, only stepping down in 2020 – doesn't believe the guides' language is too loose, but says enforcement is a major problem. Many companies the FTC has contacted over the years didn't even know the 'Green Guides' exist, she says.

In the rare instances when the commission has taken a company to court, the amount of resources this has chewed up has made it clear that it isn't viable to pursue more than a handful of companies for greenwashing. 'The FTC is a tiny agency,' says Engle. 'It has limited resources but a huge jurisdiction and remit.'

Since the explosion of social media marketing in the early 2000s, the commission has found itself even more powerless to stay on top of deceptive green marketing. 'Before the internet, it cost money to have a TV or a magazine ad. But the barriers to entry on the internet are very low,' says Engle. 'It's hard to keep up with everything.'

CHAPTER 6

PLASTICS MAKE IT POSSIBLE

A campaign to 'inoculate' consumers against the negatives

In the spring of 1993, people watching prime-time shows like *Quantum Leap*, *America's Funniest Home Videos* and *Murder, She Wrote* started seeing a series of high-budget, evocative advertisements.

A young boy is aggressively tackled on the football pitch – he slams into the ground, but his head is protected by a helmet. A grandfather bends down to hug his young granddaughter after an operation makes his pain disappear. 'Today will be a better day for a lot of people, simply because of a material we call plastics,' croons the voiceover. Another ad shows a woman undergoing chemotherapy delivered through a plastic disc. A violin plays soulfully as images of her life flash on screen. 'Miracles come in all forms,' concludes the ad. 'Sometimes plastic.'

A mother answers the door, leaving a young girl unattended, who promptly tries to stick her hand in an electrical outlet and then open a medicine bottle. Each time, plastics are in place to avoid a disaster. Yet another ad features upbeat testimony from children.

'They help protect my patella,' says a helmeted, rollerblading child clutching his knee pads. 'This vest helped save my dad's life,' exclaims a daughter, snuggling into the bulletproof vest-clad chest of her police officer dad.

There was no mention of litter or strangled birds. There was no mention of landfills or incineration. There was no mention of recycling. The spots featured items like incubators, egg cartons and helmets. They reminded viewers about plastic's durability, shatter-resistance, medical uses and ability to protect food.

*

The ads were the face of 'Plastics Make it Possible', a high-budget campaign bankrolled by the plastics industry that aimed to systematically replace Americans' antipathy towards plastic with new, positive emotions.

The seeds of the campaign had been sown in June 1992 when the CEOs of the biggest plastics makers in America were huddled together at the Greenbrier picking over the industry's plight. The stately, plush resort, sprawling across 6,500 acres in the Allegheny mountains of West Virginia, was the venue of choice each year for the annual meeting of North America's powerful chemicals industry. Anyone flying into the nearby airport that week was greeted by what locals called 'the Chemical Air Force', a long line of parked private planes from which the chino-clad, closely shaved men who headed up some of the world's richest companies disembarked, ready to play golf, smoke cigars and strike deals. Many of the men brought their wives along for the week to luxuriate in the resort's mineral baths, perambulate its manicured lawns and partake in elaborate six-course dinners.

That year, though, the mood among the gathering's plastics coterie was decidedly less buoyant than usual. Sixty-four per cent of Americans now thought the risks of plastics outweighed their benefits – a figure that had more than doubled since the *Mobro*'s unhappy voyage in 1987. The public's approval of plastics was lower than its opinion of big tobacco. The many recycling trials by companies like Dow, DuPont and Amoco, and the plastics industry's lobbyists' frequent visits to Capitol Hill, hadn't turned the tide of public opinion.

The men at the Greenbrier settled down to business, arraying themselves around an enormous oak table. On the agenda for discussion that day was a proposed $30 million-a-year ad campaign to shore up the plastics industry's reputation.

Most companies in the room were already spending millions on advertising and fighting anti-plastics legislation. Pay up for a campaign that would make no mention of their own brands? 'Not a chance' was the general refrain.

After listening to 20 minutes of bickering, Jon Huntsman – the strapping CEO of Huntsman Chemical – had had enough. He abruptly pushed back his chair and stood up, pulling a wad of money from the back pocket of his trousers and slapping it down on the table in front of him. 'Gentlemen,' he thundered as the room fell quiet. 'This is the thing to do. We all need to pay up. I own my company so my contribution is coming directly out of my own wallet. The rest of you don't and your share will come from your companies.'

Huntsman had a strong gaze, a measured manner and a reasonableness about him that was hard to quibble with. He was a 'pull yourself up by your bootstraps' kind of guy, and everyone in the room knew how he had turned himself into a billionaire despite growing up in potato country in Blackfoot, Idaho, in a family that had been too poor to afford indoor plumbing.

As a boy, Huntsman mowed lawns, washed dishes and picked potatoes for pocket money. His first brush with plastics came when he developed an egg container for his wife's uncle who ran a distribution company. Huntsman later founded the packaging company that created the McDonald's clamshell in 1974. He went on to become a major polystyrene supplier to the fast-food giant. Some people called him 'the polystyrene king'. Coming from poverty gave Huntsman a confidence – and a sense of freedom – that his peers lacked. 'If he lost it all and he had to go back to nothing, he'd have thought, *Give me six months and I'll be back on my feet again,*' remembers his son Peter.

And yet, by 1992, even Huntsman's appetite for risk was being tested by the current environment. When McDonald's switched to paper in 1990, a $200 million market in polystyrene foam containers vanished virtually overnight. Huntsman knew that much more could be lost. Something major had to be done to give the plastics industry a new licence to operate. 'The time to act is now,' he told the men assembled at the Greenbrier.

Huntsman's calls for an industry-funded campaign were backed by Dennis McKeever from Dow and Nick Pappas from DuPont, two of the highest-profile executives in the industry.

'Once those three do it, you'd have to be out of your mind to stand alone,' says Ron Yocum, then CEO of Quantum Chemical, who attended the meeting.

Many in the room were gripped by the fear that anti-plastic sentiment could wipe out billions of dollars in sales, remembers Yocum. 'We had probably collaborated before but never to the degree that we did on this. We were facing a real threat.'

*

The CEOs agreed to equally share the $30 million a year needed for an ad campaign. They tasked the Council for Solid Waste Solutions' 38-year-old public relations head, Jean Statler, with organising it.

Statler hired an agency which created a campaign called 'Take Another Look at Plastic'. Its flagship ad showed plastic containers raining down around a smartly dressed man and then reappearing as new containers. 'We were aiming to tell you about all the conveniences you get from this material, so just take another look and don't dismiss it,' she explains.

But 'Take Another Look at Plastic' didn't resonate with consumers. 'It bombed,' says Statler matter-of-factly. 'It just looked like plastic was raining on you.'

Huntsman stepped in. He told Statler she should talk to Richard Wirthlin, a fellow Mormon from Salt Lake City who for years had worked as Ronald Reagan's chief pollster and strategist.

Wirthlin began the line of work that would define his career as a side gig. Trained as an economist and statistician, he was chair of the economics department at Brigham Young University when, in 1964, he agreed to help a political scientist do some polling. Wirthlin enjoyed the work so much he took on more. Eventually, he set up his own opinion research firm, initially working from his kitchen table.

In the fall of 1968, Wirthlin was a visiting professor at Arizona State University when the phone rang. The caller on the other end identified himself as Tom Green, saying he wanted Wirthlin to conduct a public opinion study about Californians' views on policy issues and state leaders. Wirthlin pressed for more details. 'Why are you commissioning this?' he asked. 'What will it be used for?' Green was reluctant to share anything further.

A few weeks later, his study complete, Wirthlin flew to Los Angeles to present the data to his mysterious client. Green, it turned out, was in fact a well-known California political operative named Tom Reed. Sitting next to Reed in the back of an air-conditioned limousine, Wirthlin learned that his real client, the man he was now on his way to meet, was Ronald Reagan, the governor of California.

In his first few moments standing across from Reagan outside his sprawling ranch-style home, Wirthlin – dressed in a brown polyester suit – remembers feeling 'about as comfortable as a fish swimming in a bowl of wet cement'. But as the afternoon wore on and the two men talked over glasses of fruit juice in Reagan's library, to his surprise Wirthlin found he had much in common with the man he had, until now, had little regard for. Reagan and Wirthlin had both studied economics. They both grew up in middle American farm communities and placed great weight on God and family. And they both had an optimistic streak that drew other people to them. Reagan also asked probing questions and seemed genuinely interested in how Wirthlin arrived at his conclusions, not just what they were.

That meeting was the start of a 20-year business relationship and an even longer friendship. The work that Wirthlin went on to do for Reagan helped the governor become US president and stay in the White House for two terms. Over the years, the pair grew close enough that Wirthlin was sometimes called in to brief Reagan while the president and his wife Nancy lounged in their pyjamas. One year, to Wirthlin's mortification, the president asked his pollster to buy an anniversary card for Nancy. 'He almost gave me an ulcer,' Wirthlin would later write in a book memorialising his time with Reagan.

Wirthlin had a proclivity for planning. The trait, which went back to his days as a Boy Scout, was compounded by Reagan's defeat against Gerald Ford in the 1976 Republican primary. Wirthlin was convinced that Reagan had come within a whisker of winning. The episode intensified his disdain for uncertainty, propelling him to spend two and a half years creating an enormous computer system he called the 'Political Information System' or PINS. It combined quantitative, qualitative, institutional and historical data sources, helping Wirthlin predict how voters would respond to Reagan's positions. Wirthlin used PINS to decide which topics Reagan should focus on, whether his advertising should run on TV or radio, and when it should air.

Observing Reagan influenced the pollster's philosophy about the most effective way to communicate. Wirthlin summed up the future president's style as 'persuade through reason, motivate through emotion', repeating the line so often that people who worked with him remember it as Wirthlin's own mantra. He spent weeks analysing Reagan's most successful public speeches, concluding that communication is only successful at shaping how people think if it's personally relevant to their lives. He encouraged the president to craft more of his speeches in this manner.

In time, Wirthlin would be called 'the prince of pollsters' and 'pollster general' of the US. His foreign clients included British prime minister Margaret Thatcher, German chancellor Helmut Kohl and Israeli prime minister Benjamin Netanyahu.

Huntsman was a prominent Utah Republican who had served in the Nixon administration before leaving to head his chemical company. He was well aware of how pivotal Wirthlin had been to Reagan's election and re-election. Huntsman also had an in; he and Wirthlin had become friendly through the Mormon Church. In early 1992, he asked the pollster if he could help the plastics industry turn its reputation around.

Wirthlin was sure he could. He wrote a ten-page memo outlining a communications campaign to reshape how regular Americans saw plastics. At Huntsman's request, he sent it to Statler at the Council for Solid Waste Solutions, which by now had changed its name to the American Plastics Council in an attempt to distance itself from its waste problems.[1]

By 1992, APC ranked among the richest lobby groups in the US, with a yearly budget of around $60 million. It had more than two dozen member companies – primarily the US's biggest chemical makers, but also P&G, which bought so much plastic for its

---

[1] Today it exists as the American Chemistry Council.

detergents and shampoos that it was one of the largest plastic packaging users in the world.

A Jacksonville, Illinois native who kept her dark hair short and favoured suits coupled with a string of pearls, Statler had arrived at APC from the American Trucking Association, where she had fought attempts to ban trucks on highways using the tagline 'Without Trucks America Stops'. She was hard-working and ambitious, shuttling each morning from looking after her three young children at home into her small office.

Lately, though, despite the long hours she was putting in, Statler had been feeling out of her depth. The failure of 'Take Another Look at Plastic' had left her out of ideas. The ad also eventually got the APC in trouble with 11 state attorneys general, who said it misled consumers into believing plastic was widely recyclable when just 3.5 per cent of plastic was recycled. At the time, APC paid $110,000 to settle the charges and promised to be more precise when making future claims.

Wirthlin's memo to Statler laid out how he'd tackle the plastics industry's reputational challenge. To win people over, he explained, the plastics industry should link the attributes of plastics with their positive consequences and show how these helped people protect and fulfil their personal values and desires.

To understand how people really felt about issues and why they felt the way they did, Wirthlin advised conducting lengthy interviews using a technique called 'laddering', in which interviewers continually probe a response with deeper follow-ups. With a better understanding of how Americans thought about plastics, he added, the industry would be able to counter the growing tide of environmentalism by promoting positive messages that resonated with a variety of people.

Statler thought Wirthlin's memo was the most impressive strategy outline she'd ever seen. 'You ever have one of those moments where a lightbulb just goes off in your head?' she gushes. 'Any campaign could try to change the image, but what Wirthlin was saying was he could change people's behaviour.'

*

Wirthlin agreed on two main goals with Statler and the chemical giants: stopping consumers from abandoning plastics, and building positive perceptions of the material and its manufacturers. In July 1992, his firm conducted two-hour-long interviews with 86 people in four cities to uncover the thought patterns and associations they held about plastics.

The interviews kicked off by asking people to describe the basic characteristics of plastic, such as its strength, affordability or versatility. They then moved on to the benefits associated with these characteristics and, finally, to the personal values that consumers associated with the benefits, like peace of mind, contentment and personal satisfaction.

Wirthlin developed a conceptual roadmap, similar to ones he had used for Reagan. The aim was to uncover pathways in consumers' minds between plastics and the values people held dear. One example of a positive pathway went like this: 'Plastic is shatter-resistant, which provides safety for myself and my family. This makes my life less stressful and gives me the sense of being a good parent. That in turn helps me feel greater peace of mind.'

Wirthlin also dug deep to find the strongest negative pathways connected to plastics. Understanding the negatives in addition to the positives meant that APC could 'capitalise on positive associations while working to inoculate consumers against the negative ones', explained an industry document. A negative pathway might read: 'Plastic is not biodegradable nor easy to recycle and the consequence of this is that it harms the environment. That impacts my satisfaction and peace of mind since I feel irresponsible and worry about hurting my child's future.'

Wirthlin's research showed that the strongest positive pathways in people's minds involved the quality-of-life, safety and health benefits that plastics could bring. He recommended that APC leverage these benefits to come up with positive messages.

A 1995 ad that formed part of the American Plastics Council's 'Plastics Make it Possible' campaign.

The next step was deducing which positive messages would be most effective. In August 1992, Wirthlin's team began phoning 1,506 randomly picked Americans to test the credibility and impact of various messages. Respondents were asked 117 questions, including whether they agreed with statements like 'plastic products and packaging considerably improve the way we live' and 'new plastic products can help create a better future for our children'.

Wirthlin ranked potential messages by their effectiveness across pro-plastic, anti-plastic and swing groups. The eventual advertising, he advised, must be targeted to appeal to people in each group: strongly anti-plastic people only responded to messages about the medical uses of plastic, whereas swing consumers were also open to messages about how plastic keeps food fresh.

As the creative agency produced mock-ups of the new ads, Wirthlin had testers recruit consumers to view animated digital versions of the intended ads, answering questions on how these made them feel. The agency then fine-tuned the ads to eliminate anything that didn't genuinely speak to people's values.

Through all this research, Wirthlin discovered something crucial. The industry should stop overtly advertising its recycling efforts and instead seed these messages through sympathetic journalists. 'When the public learned through news articles, rather than advertising, about the plastics industry's advancements in enhancing the environment, these messages did carry credibility,' explained an industry document describing the campaign.

Accordingly, APC accelerated its outreach to journalists, emphasising the many efforts it was making to improve recycling. Simultaneously, though, it also shifted gears to tout plastics incineration.

McDonald's recycling failures had turned out to be just the tip of the iceberg. Even attempts to recycle plastic soda bottles – theoretically the most easily recyclable plastic – had run into trouble after companies discovered their recycled resin was more expensive than virgin. Quantum's milk bottle recycling plant, after years of operation, was barely breaking even.

George Rizzo, the head of polymers for Exxon, remembers visiting a chemical industry-funded plastics recycling plant in Massachusetts, only to find it was making most of its money collecting coins that people had thrown away. 'They had a vibrating feed system on a mesh screen big enough for quarters and stuff to fall through and they'd make hundreds of dollars a day on change,' he chuckles. 'They didn't make any money on the recycling itself.'

Exxon also ran into trouble with a polypropylene recycling operation it set up ostensibly to provide recycled raw material for woven fabric. 'When we started seeing what was coming in to be recycled, we recognised these businesses were probably never going to break even, let alone make money,' says Rizzo. 'There are so many dog-gone plastics and so many doggone colours. There are so many contaminants that go into it.'

By 1993, Dow and DuPont had abandoned plans to build plastics recycling plants, while other companies – like Hoechst Celanese and OxyChem – had sharply curtailed their own plans.

APC began paying to place articles and editorials by 'respected third parties' in newspapers. They described how plastic could be safely burned to produce energy.

It was the same old argument the industry had been pushing since the 1960s. Wirthlin's hard work paid off. Within a year of his campaign's launch, polling showed that Americans' feelings towards the plastics industry, measured using a 100-point feelings thermometer scale, had jumped from a 'moderate' 52 to a 'warm' 60.

By 1996, 76 per cent of Americans felt the benefits of plastics outweighed the negatives, up from 63 per cent at the campaign's inception. That year, Rutgers' Center for Plastics Recycling

Research closed its doors. Funding from the plastics industry had dwindled and towards the end it had just two staffers, down from 18 just a few years earlier. *Plastics News* reported that among the reasons for the centre's closure was 'a perceived decline in the necessity and practicality of plastics recycling'.

APC research showed that the number of 'anti-plastic' consumers had fallen by more than 50 per cent between 1992 and 1996, amounting to a shift of 10 million people. The number of bills proposed at the state level which would hurt the plastics industry had declined by half in the five years since 1991.

'Dick Wirthlin shifted the debate from "Plastics are a scourge on our earth and filling up landfills" to "You can't do without them in health, safety and medical uses and that's what keeps us safe",' says Statler. 'It really did change the narrative and the decision-making, and gave people permission to use plastics.'

CHAPTER 7

THE CHASING ARROWS

'They ran with it until the cows came home'

An odd symbol, made up of three arrows arranged in a triangle, began showing up on plastic containers across America in the fall of 1988. Inside it was a number.

The idea to put codes on plastic containers came from the Society of the Plastics Industry. By 1987, Lewis Freeman, SPI's head of government affairs, had begun hearing that the fledgling plastics recycling industry was struggling to make sense of the dozens of different types of plastics they were receiving. The plastics had different melting points and other properties, which meant they couldn't just be mixed together for recycling.[1]

---

[1] Virtually all plastics recycled today use a method called mechanical recycling – which washes, chops and melts plastic. Plastics can only be recycled a few times under this method, since each cycle degrades the plastics. Most packaging that is made from a single type of plastic can technically be recycled. Whether a plastic *is* recycled in practice boils down to economics. A plastic is likely to be recycled if a) someone is willing to pay for recycled resin, and b) the price they pay is higher than the costs of collecting, sorting and reprocessing the old plastic. Recycling is a business and so having a buyer for recycled plastic is as important as whether a plastic package is put in a recycling bin.

'Plastics is not really one material, it's umpteen materials,' explains Freeman. 'While plastics share a similar molecular structure and most are made from oil or natural gas, they're otherwise quite different from one another.'

Before he joined SPI in 1979, Freeman worked as a lobbyist for the American Petroleum Institute, fighting senator Ted Kennedy's push to break up big oil companies. At SPI, where he stayed for more than 20 years, Freeman dealt with anything that could pose a reputational risk to the plastics industry. He spent much of his time convincing companies to make changes that would forestall the risk of regulation.

When it emerged that dozens of babies each year were dying by drowning in large plastic buckets – at 5 gallons the buckets were so heavy that if an infant fell into them, they didn't tip over – Freeman was the man who rallied the industry to hand out warning stickers to parents buying the buckets. The companies, he remembers, didn't want to add permanent labels which made the buckets a few cents more expensive. Eventually, they capitulated when it became apparent their legal liability was enormous.

'Companies are essentially all the same regardless of industry,' says Freeman. 'They don't like to be told by someone else that they need to do something, period.'

Back in 1987, shortly after the *Mobro* had set sail from Long Island, Freeman took the complaints he was hearing about recycling to SPI's public affairs committee. Since the industry saw recycling as a tool to mitigate reputational damage, the public affairs group, consisting of men from big packaging makers like Owens Illinois and American Can Company, was the natural place to discuss it.

The dizzying array of plastics on the market was hardly the only issue plaguing recycling. Plastic's popularity came down to it being light, cheap, versatile and robust. But being light and cheap hurt on the other end. Hauliers, who were paid by the ton to collect recycling, made far more money filling their trucks with heavier aluminium or cardboard than with lightweight plastic. Things were worse for some plastics than others. Polystyrene foam was economically unviable because it was mostly air. Plastic bags, wraps and films also had to be collected separately or they gummed up sorting machinery.

Packaging makers preferred virgin over recycled plastic since it was better quality and usually cheaper. If there were no buyers, it didn't matter how technically recyclable something was – it wasn't going to be recycled. Back in the late 1980s, only containers made from PET (the plastic used in single-use drinks bottles) and HDPE (commonly used to make milk jugs and detergent containers) were being recycled in any significant volume.[2] These plastics weren't turned into new soda bottles or milk jugs, but instead downcycled into lower-grade construction material that was just one step removed from the landfill. All the other kinds of plastics went straight to landfills or incinerators, if they weren't littered.

Environmentally, the two main reasons to recycle are to save raw materials like fossil fuels (as well as water and energy) and to cut down on landfilling, incinerating, dumping and littering. Recycling can save between 30 per cent and 80 per cent on carbon emissions compared to making virgin plastics. But plastics that have to be transported long distances and cleaned with lots of water, or that require very high temperatures to melt, are often not worth recycling, either environmentally or economically.

Slapping a code on the bottom of plastic containers wouldn't fix most of these problems. But at least it would help recyclers know what they were dealing with, Freeman told SPI's public affairs committee. Many plastic resin producers in the room were against

---

[2] The situation has not changed today: PET and HDPE bottles continue to be the only two plastics recycled at significant scale in the US and across much of the world. Non-bottle items, like trays, are rarely recycled. More on recycling in the FAQs.

the idea. They feared that including a code would encourage consumer goods makers to spurn plastics that weren't being recycled. Even the makers of recyclable PET and HDPE containers didn't embrace Freeman's proposal. Freeman compares them to the bucket makers who preferred to sit on their hands until they had a legislative gun pointing at their heads. 'The bottle manufacturers opposed it because it required them to do something,' he says.

Freeman eventually prevailed. He insisted the code was a way to forestall mandatory regulation that could be far more expensive and onerous. For plastics that weren't currently being recycled, the code was the first step towards enabling this, he added, since it meant they could be more easily sorted.

And so the 'resin identification code', as the industry called it, was created in 1988. While there were dozens of different types and subtypes of plastics, SPI – looking to keep costs and complexity low – grouped them into seven broad categories, which still stand today. They are:

Polyethylene terephthalate (PET), used for soda and water bottles

High-density polyethylene (HDPE), used for milk jugs, detergent containers and shopping bags

Polyvinyl chloride (PVC), used for credit cards and pill packs

Low-density polyethylene (LDPE), used for disposable gloves, trash bags and dry-cleaning bags

Polypropylene (PP), used for yogurt tubs, takeaway boxes and butter containers

Polystyrene (PS): the solid kind is used to make disposable cutlery and cups, while the expanded kind (EPS) is used for foam egg cartons, meat trays and fast-food containers

Other plastics: a catch-all for remaining plastics including multilayer packages like pet food pouches and ketchup sachets that incorporate different types of plastic, as well as bioplastics.

*

To separate the number from other descriptors used on containers, SPI enclosed it in the chasing arrows symbol. It was a strange choice, one that would cast doubts over the plastics industry's motives for decades to come.

Back in 1970, Gary Anderson, a 23-year-old architecture student at the University of Southern California, had seen an enormous wall-sized poster advertising a design competition. Sponsored by Container Corp, a paper packaging maker that was also the largest paper recycler in the US, the competition required participants to design a symbol 'for the love of earth' to 'symbolise the recycling process'.

Anderson's design – featuring three arrows twisting and returning into themselves – won. He got a $2,500 tuition grant and a trip to Chicago in September 1970 to attend a press conference at Container Corp's headquarters. 'I was kind of an arrogant little punk student and I thought the whole thing was kind of silly, actually,' recalls Anderson, who back then sported a goatee and wore his red hair – bleached blond by the California sun – in curtains parted slightly to the side.

Through the 1960s, the paper industry – much like plastics would later – had faced mounting criticism about how its disposable products were flowing to landfills. Container Corp made the new chasing arrows symbol available to the entire paper industry for use on shipping containers and folding cartons, saying it hoped the symbol would spread awareness about the importance of paper recycling.

'It was a marketing tool,' explains Anderson.

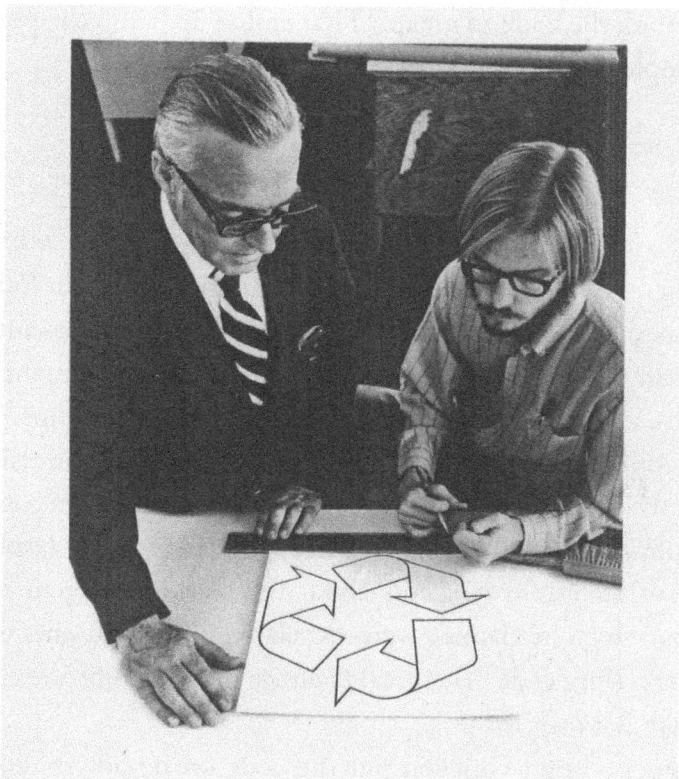

Gary Anderson (*right*) is shown alongside the chasing arrows symbol he developed in 1970.

Despite this, in 1988, when the Society of the Plastics Industry decided to use the chasing arrows on plastic containers, its executives insisted the resin identification code was not meant to indicate recyclability. It also said the code was not aimed at consumers.

Freeman says SPI chose the chasing arrows to distinguish the numbers from any others that might be found on containers, and that it was only meant to help recyclers sort plastic resins from one another. 'It was not an attempt to deceive people that because an item had the code on it, it was recyclable,' he says. But, looking back, Freeman acknowledges that recyclability is exactly what people took the code to mean. 'That ended up being the presumption people drew – and still draw until this day.'

*

Within a few months of its inception in 1988, the SPI code[3] began catching on across the US. Colgate put it on its bottles for Palmolive and Ajax dishwashing liquids. P&G slapped it on Jif peanut butter jars, bottles of Crisco oil, Tide and Cheer laundry detergent bottles and tubs, and even on its plastic detergent measuring cups.

Including the chasing arrows symbol together with the resin identification code on products that couldn't be recycled gave consumers the impression that they could. 'They are made from polystyrene,' a P&G executive told reporters about the plastic detergent measuring cups, which he claimed were recyclable. 'That's number 6 on the plastic recycling code.' But local facilities didn't accept the cups and they were not recycled.

Kentucky Fried Chicken put the code on its polystyrene foam containers. It said it expected polystyrene foam to be recycled in the future. H.J. Heinz Co. put the code on its ketchup bottles, which contained six layers of polypropylene and film barriers attached by an adhesive. The polypropylene couldn't be recycled because the various barriers couldn't be separated from it. A Heinz spokesman batted away a reporter's questions about recyclability, saying consumers were more worried about the ketchup's freshness.

State officials supported the code, hailing it as a game-changer for recycling. 'Container coding makes plastic recycling practical, pragmatic, makes it good business,' Texas's land commissioner told reporters in 1989. So upbeat was he about the code's impact on

---

[3] SPI's resin identification code is made up of a number surrounded by the chasing arrows symbol. I refer to it as the code throughout.

plastics recycling that he predicted the price of used plastics would soon rival that of aluminium.

By the early 1990s, at the urging of SPI, 39 states had enshrined the code as law on rigid plastic containers. Companies eagerly embraced the law, but also started putting the code on flexible plastic wrappers for everything from pantyhose to Subway sandwiches.

Some brands had begun to use the exhortation 'Please Recycle' alongside the chasing arrows symbol on plastic products and packaging that couldn't be recycled, claiming this was an educational effort. Surveys showed that the majority of consumers thought that 'Please Recycle' meant consumers could recycle those products in all or most communities in the US.

The Federal Trade Commission's 'Green Guides' didn't explicitly prohibit anyone from making general statements purportedly intended to encourage people to recycle. It had also said that the chasing arrows wasn't a claim to recyclability as long as the symbol wasn't used prominently. Increasingly though, companies *were* using it prominently. They were putting the symbol on the sides and fronts of packaging.

'Over time, even companies who initially opposed developing the code grabbed on to it and started putting it on everything,' says Freeman. 'Companies decided it was in their interest to look green and they ran with it. They ran with it until the cows came home.'

*

By 1992, it was clear that SPI's code, intended to help plastics recycling, was in fact hurting it.

As more Americans relied on the code to make their recycling decisions, recyclers saw a surge in the quantity and variety of plastic containers coming their way. Given that only PET and HDPE bottles were being recycled, the recyclers had to sort all the other plastics out and pay to have them landfilled or burned. They also had to deal with irate consumers who believed that if a product carried the chasing arrows symbol, it must definitely be recyclable.

'The consumer was blaming us, saying we didn't know how to do our jobs, didn't know what's recyclable,' remembers Coy Smith, who worked as a recycler in San Diego. 'They'd say, "What do you mean it's not recyclable? It's got this triangle thing on it!"'

The code, it turned out, was also pretty much useless to recyclers. The numbers only denoted broad groups of plastic. But the groups contained an array of subtypes that couldn't be recycled together. Blow-moulded HDPE used for bottles can't be recycled with injection-moulded HDPE used for other products. The two types have different melt flow indexes – which means one isn't as viscous as the other, a discrepancy that can cause defects in final products. Yet both were labelled as a number 2 plastic. Similarly, PET bottles can only be recycled with a limited number of tubs and trays. The tubs and trays are more brittle – too many of these can make the recycled plastic unsuitable for holding carbonated drinks. Despite this, all PET was labelled with the number 1.

'One through seven doesn't cut it,' says Smith. 'Those are major, major headline categories and there are hundreds of subtypes with different properties.' As a board member of the National Recycling Coalition, a major trade body for US recyclers, Smith had links with recyclers and environmentalists across the country. In 1992, he began rallying support for 'Take the Wrap', a campaign aimed at stopping the plastics industry from using the chasing arrows symbol on packaging that wasn't recycled.

One of Smith's biggest allies was Athena Bradley, a 33-year-old who administered a recycling programme in Chula Vista, California. One afternoon, Bradley answered the phone at work, only to be called 'a stupid idiot' by the woman on the other end.

Her caller had just learned that her plastic containers weren't being accepted for recycling, despite bearing the code.

Bradley, like Smith, didn't buy SPI's claim that the codes on the bottoms of containers were intended to help recyclers. 'It's ridiculous. You can't have recycling coming down the line and people individually picking up bottles looking for the 1's and 2's,' she scoffs. 'It was such a marketing campaign and, as recyclers on the ground, we were absolutely trapped. We'd tell residents this isn't recyclable. But they'd say, "It has the arrows on it. Of course it's recyclable."'

Bradley joined Smith in galvanising other recyclers and consumers to mail used plastic containers to SPI as a way to pressure the trade body to stop using the code. The 'Take the Wrap' campaign organisers also approached state attorneys general like Skip Humphrey, demanding that they crack down on companies using the code on products that weren't being recycled.

With a backlash building, in February 1993, SPI agreed to open discussions with the National Recycling Coalition over the code's future. Over 15 months, a working group made up of 14 people, half from NRC and half from SPI, met regularly to hash out recommendations on how to improve the code. Their goal was to reach a consensus over what a new code might look like, which could then be voted upon by the wider membership of both associations.

Coy Smith – a talkative man with sandy hair and a wide forehead – was one of NRC's seven representatives. He proposed replacing the chasing arrows with a square. The shape, he said, would still differentiate the code from other markings on plastic containers, but consumers wouldn't conflate it with recyclability. The plastics contingent was dead against making any such changes. Swapping the chasing arrows for a square could cost up to $80 million. Machinery would need to be changed, existing moulds would have to be thrown out and new ones bought.

In the face of the plastics makers' recalcitrance, negotiations looked like they might grind to a halt. The other NRC representatives in the group began leaning towards striking a compromise: replacing the chasing arrows with a triangle that had no arrows. The shift would cost far less than what Smith was proposing, since mould makers needed only to modify, rather than replace, existing equipment to block out the arrowheads and produce a triangle. Under this plan, the numbers would be supplemented by a letter that indicated the subtype of plastic resin based on characteristics like its melt flow.

The six other NRC representatives in the working group all backed the compromise. Smith thought it was utter garbage. 'My stance the whole time was that the symbol, if it's going to change, can't be a triangle because people are already identifying a triangle as meaning it's recyclable,' he says.

In the weeks leading up to the group's final recommendations, Smith had clashed repeatedly with NRC's president, Mark Lichtenstein, and its former president, Terry Guerin. Both men thought swapping the chasing arrows for a triangle was a big step in the right direction. 'Some people felt like it was dancing with the devil, even sitting down with the plastics industry, but I didn't feel that way,' Lichtenstein says. 'This was a huge coming-to-the-middle by the plastics industry. Was it perfect? No. But you don't get everything you want in one shot.'

Once it became clear that Guerin and Lichtenstein were immovable, Smith began working the phones to get other California-based recyclers – of which there were by now a fair few – on his side. Many, like him, were fed up with getting plastic packaging they couldn't recycle and mistrustful of any compromise the plastics industry offered.

On the morning of 21st May, 1994, more than two dozen recyclers who served on NRC's sprawling board assembled to vote on the proposed overhaul of the chasing arrows symbol. SPI's board was scheduled to meet to cast their own vote five days later. If the majority of both boards approved, the process of asking the 39 states that had adopted the code to change their laws to replace the chasing arrows with the triangle and more nuanced numbers would begin.

Guerin kicked off the meeting feeling confident. Fifteen months of negotiations had, he told the room, culminated in something acceptable to both sides. The new symbol would end the use of the chasing arrows that consumers found so confusing and adapt the code to be more useful. Both sides should be congratulated on the cooperation they had shown, a beaming Guerin told the assembled board members as Lichtenstein, seated by his side, nodded approvingly.

But the new symbol was voted down. 'There was a coup,' says Lichtenstein, who still harbours bitter feelings about the episode. 'We were so shocked and surprised. We felt like we had mud on our faces.'

After the vote, Lichtenstein was so incensed that he made the NRC board break with protocol, calling a second vote. That failed too and with it the fragile coming-together of the plastics industry and the recyclers fractured irreparably. Within the recycling community, the episode created a rift that Lichtenstein says can't be bridged. 'It drove wedges between friendships and relationships and created bad blood that still runs today,' he says.

As for SPI, within months the trade body had decided to double down on the existing code – chasing arrows and all – expanding its scope well beyond the original intention. SPI's president Larry Thomas told members that the body now condoned the use of the code on plastic packaging beyond rigid containers as a way to encourage recycling.

More than two decades later, the chasing arrows symbol is still widely used on plastic packaging. In 2021, California banned it from being used on products or packaging unless the items are regularly collected and processed for recycling in the state. But California is an outlier. The chasing arrows, and the numbers inside, continue to show up on plastic packaging not just across the US but also globally. No enforcement action has been brought by the Federal Trade Commission to date against any company over the use of the code.

Today, some 35 years after the first plastics recycling programmes were rolled out in the US, plastics recycling has largely been a failure. Only PET and HDPE containers (numbers 1 and 2) are recycled at scale. Even their rates are poor, with only 28 per cent of bottles made from both plastics collected for recycling. The overall recycling rate for all plastic packaging that American consumers throw away is 13.6 per cent.

'I started addressing the issue a third of a century ago and we're still dealing with some of the same issues,' says Lewis Freeman. 'We haven't solved them as yet.'

About 92 per cent of Americans polled about the resin identification code in 2019 by the Consumer Brands Association, a trade body, said they didn't understand it. Nearly 70 per cent said they assumed any products with the chasing arrows and numbers on them would be recyclable.

As for Gary Anderson, he's saddened that the symbol he designed all those decades ago has become so closely associated with greenwashing. 'I've heard some very hateful things floating around out there about it,' he says. 'I still believe it is useful. But I feel sorry for the people who are rather misinformed about it.'

CHAPTER 8

THE GARBOLOGIST

'Much of plastic's bad reputation is undeserved'

William Rathje, the garbologist, paused dramatically as a photo of a newspaper from the 1960s carrying the headline 'Apollo Orbits Moon' flashed on the big screen behind him.

It was the last day of January in 1994. Rathje had flown to Park City, Utah, to address Huntsman Chemical's customers and suppliers. For many attendees, the company's annual customer seminar – a ten-day extravaganza hosted at a luxurious ski resort – doubled as a vacation.

Rathje's fame had grown substantially since the late 1980s when McDonald's and Don Shea at the Council for Solid Waste Solutions had hired him to tell lawmakers about how nothing degrades in landfills. While the ongoing 'Plastics Make it Possible' campaign was aimed at winning the hearts and minds of regular Americans, companies had begun leaning on Rathje's findings to convince legislators that alternatives to plastic wouldn't help solve the problem of landfills filling up.

The University of Arizona anthropology professor was increasingly in the limelight, much to the envy of his peers, many of whom regarded him as quirky and his obsession with America's discards as baffling. By now, Rathje counted Procter & Gamble and Pepsi's snack business Frito-Lay among his funders. As his relationships with big corporations had flourished, the professor had become increasingly vocal in his defence of the plastics industry and its customers.

'Bill was fairly conservative and I think that coloured his views and shaped where he went for support,' says Michael Brian Schiffer, a fellow archaeologist who worked closely with Rathje at the University of Arizona. 'He got a lot of support from industry.'

At the 1994 talk at Huntsman's retreat, Rathje spoke wearing his customary blue jeans and a tie bearing a pull-tab clip – the kind used to open soda cans. He told his audience that the perfectly legible newspaper bearing the Apollo news was one of hundreds of such finds his team had discovered intact in landfills over the past few years. There was also a whole T-bone steak fully cooked and wrapped in a paper towel, a ceramic mug, a sock, hot dogs, grass clippings and guacamole that Rathje estimated was thrown away in 1967. He dated his findings by using nearby newspapers and phone books excavated as part of a single layer of trash. 'Dig a trench through a landfill and you will see layers of phone books, like geological strata, or layers of cake,' he once wrote in an article explaining his methods.

While food and yard waste did degrade slowly – about 25 per cent in the first 15 years – it then mummified with little to no change for another 40 years at least, according to Rathje. Everything else, he said, retained its original weight, volume and form over this period. Mummified, fully intact waste had turned up in landfills in arid climates but also in more humid parts of the country. 'The dynamics of a modern landfill are very nearly the opposite of what most people think,' he wrote in his 1992 book, *Rubbish!*

The Archaeology of Garbage. 'They are not vast composters; rather they are vast mummifiers.'

Rathje's findings were a real shot in the arm for the plastics industry. One of the biggest misconceptions companies were battling was the idea that plastics were filling up landfills while paper and food harmlessly degraded.

'It was a big hit and it gave the industry hope,' remembers Don Olsen. Each year, the Huntsman communications head collected Rathje from the airport and took him up to the resort for his highly anticipated speech. The 1994 talk was the third in four years that the dishevelled professor had given to Huntsman's customers.

By now, Rathje's findings about all the non-plastic discards that hung around in landfills had repeatedly been used to weaken Steven Englebright's landmark bill to ban plastic in Suffolk County. Originally slated to be implemented in July 1989, nearly five years later the bill's fate was still hanging in the balance.

The many delays had started in the summer of 1988. That's when the Society of the Plastics Industry sued Suffolk County, alleging that the plastics ban was illegal. Among SPI's arguments was that using paper rather than plastic would use up *more* fossil fuels since trash incinerators would burn less efficiently in the absence of plastics, which raised temperatures. The lawsuit also said that paper was bulkier and would need more trucks to transport. A judge ordered an environmental impact study, putting the law on hold. Then the New York Court of Appeals ruled that SPI didn't have the legal standing to challenge the bill.

From here on, lawmakers sympathetic to the plastics industry and to the plight of retailers proposed various amendments and exceptions to the bill. SPI parachuted its most formidable lobbyists into Suffolk to warn of the disaster that would ensue should any plastics be banned. They also demanded repeated delays to allow SPI time to demonstrate that polystyrene could be recycled.

The trade body flew in Rathje to testify that replacing plastic with paper would not lessen the amount of trash flowing to landfills. The garbologist's stance infuriated Englebright. He saw Rathje's arguments about how paper and food weren't degrading in landfills as irrelevant, since paper could be recycled and food waste composted – and, anyway, the county's landfills were slated to be shuttered. Rathje, he says, ignored this and 'purposefully used his background in science in order to earn a lot of money from the plastics industry'.

'It was pretty disingenuous,' says Englebright, whose voice, even today, thickens with anger as he recalls Rathje's testimony. 'He was a man of science, he understood the issue and he basically made a false presentation.'

*

Rathje was born in South Bend, Indiana. He had harboured an interest in archaeology since he was a child, falling in love with the profession after his parents gave him the *Golden Adventure Book of Archaeology*. He studied classic Mayan burial practices, earning a doctorate from Harvard University, and was well on his way to becoming one of the most preeminent Mayanists of his time when fate got in the way.

In 1971, his first year teaching at the University of Arizona, a 26-year-old Rathje asked his students to examine what artefacts might say about patterns of behaviour. Two of Rathje's students collected garbage from various neighbourhoods in Tucson to examine divergences between stereotypes and reality. The discards yielded some counterintuitive finds. Rich and poor households ate the same amount of steak and drank the same amount of milk. Poor households bought more expensive child-education items and household cleaners.

Collecting and examining garbage caught on among other students and, two years later, Rathje entered into an arrangement with the City of Tucson to get deliveries four days a week of randomly selected household trash. 'The Garbage Project' was born. The project's logo, featured prominently on Rathje's business card, was a trash can inscribed with the phrase 'Le Projét du Garbage'.

Rathje saw his work as no different to regular archaeology, except that he was looking at modern trash to learn about human beings today. His creation of what was for all intents and purposes a new field of study was so widely acknowledged that the *Oxford English Dictionary* in 1975 credited him with coining the term 'garbology'.

Back in the early 1970s, Rathje was a slender man with a mop of dark hair so thick that it resembled the plastic hairpieces Lego characters wear. He had a laugh that shook the room and a range of quirky hobbies, including photographing clouds shaped like animals and making elaborate food sculptures from watermelons, potatoes and carrots. Friends remember his ability to guzzle a six-pack of beer with no apparent effort, along with his delight in spirited debates about even the most trivial topics. He often greeted perplexed students with lines like 'Ding Hoy Buckaroo!' A reporter for the *Boston Globe* once described Rathje as 'one of those people who loves what he does so much that he projects electricity'.

For years, Rathje led the university's archaeological studies of modern refuse, spending much of his time examining garbage. Archaeologists regularly gossiped behind his back, saying he was throwing his Harvard degree away by mucking about with trash. Given that there was no digging involved, he wasn't a 'proper' archaeologist, they sniffed.

In April 1987, largely to thumb his nose at his many critics, Rathje kicked things up a notch. He began excavating a landfill in Tucson. Wearing a safari-patterned T-shirt and Aviator sunglasses, he supervised a large, graphite-toothed digging bucket attached to an extendable pole that plunged into garbage. The bucket was like a cookie-cutter with a bottom that closed once it arrived at its desired location.

Rathje wanted to cross-validate the data he was getting from fresh garbage with what was being landfilled. He also wanted to know what happened to trash after it was buried. Once he was actually digging things up, Rathje began fashioning himself as a modern-day Indiana Jones. In addition to his pull-tab tie clip, he often wore a belt buckle shaped like a garbage truck and a garbage-pail lapel pin.

Bill Rathje, pictured on a garbage barge in 1991 heading to Fresh Kills landfill on Staten Island. © Louie Psihoyos

His timing couldn't have been better. A month after the *Mobro 4000* had left Long Island, his team lowered its first bucket into the Vincent H. Mullins landfill in Tucson. As the nation's concerns about garbage hit fever pitch, Rathje became a hero to many in the plastics industry. He himself liked to say that he brought hard science and cold facts to conversations that were too often based on wildly inaccurate assumptions.

'Bill Rathje showed that nothing in landfills biodegrades, not paper, not food waste, not anything,' says Olsen, who travelled across the country to present Rathje's findings to dozens of groups considering bans. 'That became very, very helpful to the industry in pushing back on this biodegradable argument.'

Rathje went on to conduct digs in landfills in Chicago, San Francisco and New York. By 1992, he had excavated 15 landfills. He once tried to set up a garbage museum to feature his more unusual findings – which included a diamond ring and a set of antique soldiers – but his plans were thrown into disarray after someone accidentally threw away his carefully sorted exhibits. In later years, when Rathje moved to teach at Stanford University, he reportedly preached about the recreational potential of landfills, including a baffling proposal to create the US's first garbage rodeo.

His digs showed that Americans were consistently overestimating the volume of disposable diapers and fast-food containers in landfills. Newspapers alone took up between 14 per cent and 18 per cent of landfill space, said Rathje. Office paper and phone books were also two big culprits. Paper overall occupied a whopping 40–55 per cent.

Englebright believed that a big reason why newspapers were flowing to landfills rather than being recycled was that discarded newspapers were bagged in plastic, contaminating the recycling process. If plastic bags were banned, newspaper and food waste would be bagged in biodegradable paper bags, he said, making them easy to recycle and compost.

But given that Rathje and his team alone were plunging into land-fills emerging with armfuls of trash and data, it was hard to quibble with the garbologist's finds. 'What we're trying to do is provide policy planners and decision makers with hard quantitative evidence about what's in garbage so they can make decisions that will really accomplish what they want to accomplish,' Rathje explained in a 1990 interview.

By the mid-1990s, it would also become apparent that the US was not in fact running out of landfill space. It was true that thousands of landfills (4,727 between 1988 and 1995) were closed, many because they didn't meet new safety regulations. But these were, on average, rather small. The landfills that stayed open – because they could afford the liability insurance and the technology to control run-offs and monitor the land – were much larger. Many were pouring money into expanding.

Why, then, had so many people believed America was running out of landfill space? One theory was that waste management companies had purposely perpetuated the idea of a shortage, since it allowed them to jack up landfill disposal fees.

Trash did need to be transported from space-constrained states in the northeast to elsewhere in the US, which added to rising garbage disposal fees. Between 1984 and 1994, landfill tipping fees surged 177 per cent. But, overall, landfill space remained plentiful. Americans could keep buying things and throwing them away, seemingly with few major consequences.

Soon after the *Mobro*'s voyage, Rathje went out of his way to tell the American public that he had no concerns about the country's capacity to bury garbage. 'I am not worried that even if present trends continue, we will be buried in our garbage,' he wrote in 1989. 'To a considerable extent we will keep doing what other civilizations have done: rising above our garbage.'

The plastics industry latched on to his findings as proof that there was no need for panic. 'It was exciting to the plastics people that there was somebody with the credentials who could speak to the other side of this – an independent source,' says Olsen.

But how independent Rathje was is unclear. His landfill digs cost $250 an hour and he was constantly looking for grants and other funding from a variety of actors, including private companies. 'He was an opportunist when it came to funding,' says Schiffer, his former colleague. 'He would get funding anywhere he could.'

Rathje passed away in 2012. The University of Arizona has kept no records of where his funding came from and people who knew him say he didn't really discuss it. 'Because the Garbage Project required ongoing funding, Rathje spent a great deal of time writing grant proposals to government agencies and soliciting contracts from businesses,' wrote Schiffer in an anthropology journal in 2015. Rathje, he added, was 'very successful' in his fundraising efforts.

In a 1993 newspaper interview, Rathje denied allegations that his research might favour his funders, pointing out that some of his funding had come from the paper industry whose products, he had found, were clogging up landfills.

What is clear is that Rathje's opining stretched well beyond his stated area of expertise, which was archaeology and how artefacts found in landfills were emblematic of behaviour. In hundreds of interviews and news articles, he made the case for everything from the causes of diaper rash to the environmental safety of landfills. He maintained that, if properly lined with clay, landfills would indefinitely contain trash without polluting underground water.

An environmental engineer pointed out that Rathje didn't have the background to speak knowledgeably about landfill pollution. 'He's out advocating things that he doesn't know anything about – it would be like me trying to do something in archaeology,' the engineer told the *Richmond Times-Dispatch* in 1998, adding that present-day landfills at best postpone when groundwater pollution occurs.

Given America's newfound fascination with trash, the professor was in high demand to be a keynote speaker. In time, he held forth to audiences not just in the US but also in London, Amsterdam, Toronto and Sydney. Of his 33-page resume circa 2006, nine pages were dedicated to recording the many talks he had given to companies, trade associations and government bodies among others. Companies who hired him included Coca-Cola, Campbell's, General Mills, Heinz, Kellogg's, McDonald's, Oscar-Meyer, Pillsbury and, of course, Huntsman.

'He was able to take a crowd of people he had talked to the year before and just hold them in the palm of his hand,' recalls Jon Huntsman's son Peter, who heard Rathje speak each year at the chemical company's customer seminars. 'He was an incredible speaker.'

The garbologist became so used to being wined and dined by America's wealthiest companies that, at a lunch in Tucson with P&G's Scott Stewart, he ordered the priciest steak on the menu, had two bites and then waved the waiter over to have it wrapped up to go.

'I said, "Dr Rathje, I guess you've left your appetite behind",' recalls Stewart, who had flown in from Cincinnati to follow Rathje around while he did a landfill dig. 'He said, "I have this German shepherd at home. He'll love this!" He had got used to people coming over from the private sector and lost a bit of humility, I guess.'

Rathje developed a particularly close relationship with P&G, which was determined to keep growing sales of its disposable diapers. P&G's environmental microbiologist accompanied Rathje on digs. Large amounts of the waste from landfills that the garbologist was excavating were sent to P&G's labs to be analysed. Against this backdrop, it's perhaps unsurprising that Rathje – despite having no kids and having likely never changed a diaper himself – would become a strong advocate for disposable diapers. In addition to espousing P&G's arguments about how cloth diapers gave babies more rashes, he also backed the company's view that cloth diapers used more energy and water. Ignoring parents who said they found diaper services convenient, Rathje argued that mothers switching to cloth from disposables would inevitably be hugely burdened.

At the annual meeting of Canadian plastics manufacturers in May 1990, Rathje told the room that studies had found that the material to make disposable diapers was so absorbent that it readily sucked up toxic leachate. 'We conclude that the solution to the problem of toxic leachate in landfills is to line them with disposable diapers,' he said. 'So if you've got kids at home and they're in diapers, you're doing your bit if you're using disposables.' His audience applauded wildly, rushing to ask for his autograph at the session's end. His remarks were picked up by Canadian newspapers – the *Globe and Mail* even put them on its front page.

Unsurprisingly, his stance sparked an outcry among many readers and environmentalists. Rathje backtracked, saying that he was only joking and would try to be more serious from here on.

Rathje's overarching message was that if something is convenient and small-scale compared to the construction debris, newspapers and phone books in landfills, consumers should feel free to keep using it guilt-free. 'Let's deal with the big-ticket items before we ask millions of mothers to torture themselves,' he told the *New York Times* in 1992. 'There are so many ways we are wasteful in this country that are not at the absolute core of modern American behaviour. Let's start there.'

Diapers, he contended, made up just 1.2 per cent of the waste in landfills. This was far lower than the estimate from Carl Lehrburger, whose scathing report in 1988 had triggered the backlash against disposable diapers.[1] Environmentalists thought diapers

---

[1] Lehrburger incidentally interviewed Rathje to write his own diaper report, published in December 1988, relying on the garbologist to conclude that 'few if any diapers had been emptied in the toilet'.

were nonetheless a problem. 'One percent of billions of tons is worth worrying about,' a scientist at the Environmental Defense Fund told the *New York Times*. 'And if we don't think about how to address that one percent, which one percent will we address?'

That same scientist, Richard Denison – who had also worked to pressure McDonald's to cut waste – said that nobody could make proper assessments of waste by looking only at landfills. 'These are renewable resources that are used once,' he said. 'We can do better than that, even with disposable diapers.'

But Rathje's lens was trained on landfills. He took a similar stance on polystyrene foam containers for fast food as he had on diapers. Food and drinks needed to be delivered in something, so why not plastic? His arguments had broad resonance. Legislators and even some environmentalists changed their tune, moving from rallying behind bans to supporting recycling.

At early hearings for Steven Englebright's bill in 1987, the Society of the Plastics Industry's chief lobbyist had insisted that plastics were no worse than paper because *no* fast-food packaging, of *any* material, could be recycled. 'The economics of fast food recycling are non-existent both for plastic and for paper,' said the lobbyist Roger Bernstein. 'There is no municipality in the United States that's recycling fast-food trash.'

But as the hearings dragged on, it looked like paper had an advantage. Lawmakers contended that paper containers could bio-degrade and, if incinerated, burned more cleanly than plastics.

And so the plastics industry changed its tune. Plastic fast-food containers, they said, *could* in fact be recycled. Plastics industry lobbyists pressed for exemptions for businesses or even entire towns that offered polystyrene recycling facilities.

Englebright argued that recycling didn't make any sense for poly-styrene foam containers. Many were taken to go and hence dumped in general waste bins. Even if put in designated recycling bins, they were too light to be collected economically. He also maintained that it was illogical to spend time and money developing recycling programmes to justify a product used for a few minutes at most. A ban, he maintained, was still the way to go.

But, by now, the US was in the throes of a recession and in Suffolk County – the site of Englebright's bill to ban plastics – worries about jump-starting economic growth were quickly surpassing environmental concerns. Polystyrene had a clear cost advantage. Foam trays cost a third of the price of paper ones, while 6-ounce foam cups cost less than half of paper ones.

At a hearing for Englebright's bill, Robert Barrett, Mobil's general manager, testified that replacing foam packaging with paper would cost Suffolk consumers $4.1 million a year. Mobil also ran newspaper ads saying, 'Foam fast-food containers: the scapegoat, not the problem'. The ads explained that 'there are no villains, and we're all "guilty". Every household, every business, every office – indeed, every American – contributes to the refuse stream every day. To zero in on the fast-food business, or the plastics industry, is to engage in scapegoating, not problem-solving.'

Kevin Dietly, an environmental consultant regularly hired by the industry, offered up an even higher estimate than Barrett's, saying the plastics ban would cost consumers $17.9 million.

The arguments easily won over legislators. On 5th April, 1994 – one month before negotiations to amend the chasing arrows collapsed – Suffolk's county executive signed into law an amendment sponsored by Republican Thomas Finlay that essentially killed Englebright's bill.

Instead of a ban, places that sold polystyrene containers and plastic bags only had to offer facilities for people to recycle these. The new law established no minimum level of actual reprocessing. That meant an establishment could comply just by providing a recycling bin and a retailer could comply by allowing customers to return plastic bags to stores or simply by offering them the choice between a paper or plastic bag.

'The word "recycling" is the answer to all our environmental concerns,' said Finlay airily.

*

Rathje's arguments didn't just help thwart proposed bans on poly-styrene – they were a boon for plastics makers overall. They lent credibility to arguments about 'source reduction'[2] that continue to be the industry's go-to defence today.

Rathje's research showed that, although the number of individual, disposable plastic containers thrown away had doubled over the previous two decades, these took up the same amount of space in landfills as a decade earlier – 16 per cent – because manufacturers had shifted to thinner, more flexible materials that were easily crushed. 'Much of plastic's bad reputation is undeserved,' Rathje wrote in a 1989 article in the *Atlantic*. 'Because plastic bottles take up so much room in our kitchen trash cans, we infer that they take up a lot of room in landfills. In fact, by the time garbage has been compressed in garbage trucks (which exert a pressure of up to fifty pounds per square inch on their loads) and buried for a year or two under tons of refuse, anything plastic has been squashed flat.'

It's worth noting again that Rathje's claim to fame centred around landfills: what went into them, what people's discards said about human behaviour, and where policymakers should focus if they wanted to save on landfill space.

'His shtick, if you will, was always about landfill diversion,' explains Robert Lilienfeld, who co-authored a book with Rathje

---

[2] The source reduction argument advocates eliminating waste at the source or reducing it by using less of a material.

called *Use Less Stuff*, published in 1998. The book was predominantly written by Lilienfeld; Rathje provided the historical information and data around what people throw away. Confusingly, despite the title, Rathje did not advocate that people should stop to consider whether they really needed something, could consume less or even perhaps do without. Rathje didn't believe parents should switch to cloth diapers or that fast-food chains and coffee shops should use washable plates and mugs to serve people eating in, even if these measures did cut waste. 'Source reduction doesn't diminish the volume of garbage by all that much and it eats away at the conveniences that lie at the heart of our lifestyles,' he wrote in *Rubbish! The Archaeology of Garbage*.

Rathje was also not in favour of regulation that would require companies to cut down on packaging, noting that it had health and convenience benefits, and that companies were anyway propelled by market forces to slash packaging to save money. He didn't believe that goals to reduce waste should trump the specific goals of retailers – such as discouraging shoplifting by packaging small items like can openers in unnecessarily large packaging. 'Who is to say when packaging is excessive?' he wrote in the *Atlantic*. 'The utility of legislated source reduction is in many respects an illusion.'

I found it difficult to wrap my head around why Rathje agreed to author a book with Lilienfeld that said people should use less stuff if he didn't in fact believe people should use less stuff. The answer became clearer once I spoke again with Lilienfeld and read more of Rathje's writing.

Rathje did in fact promote a certain type of source reduction. However, this didn't necessitate people compromise their convenience-focused lifestyles. Instead, he suggested they could slash waste by shifting from heavier, more rigid bottles and jars to lighter, more flexible pouches, wraps and bags, cutting down enormously on volumes flowing to landfills. By this measure, he argued, plastics were a great tool to *reduce* waste.

Unsurprisingly, this argument found a keen audience. Between October 1993 and the end of 1998, Rathje spoke at more than 20 different seminars organised by the Flexible Packaging Association. The trade body represented more than 200 companies making crushable packaging, such as bags, wraps, pouches and sachets, much of it plastic.

FPA's audience at these seminars largely consisted of teachers who were sometimes taken on a tour of a plastics production plant afterwards. The teachers were encouraged to accept educational materials and to spread the message in their classrooms that creating less waste was easy: for instance, one could choose a plastic pouch over a heavier steel can. The seminars reached more than 12,000 teachers responsible for more than 1.6 million students.

Flexible plastics – particularly ones made from a blend of different materials or different types of plastic – were very difficult to recycle, but Rathje wasn't preoccupied by this, as he once told *Plastics News* bluntly. 'The value of plastics is in the source reduction that it generates in holding products and not in how much of it can be recycled.'

In the 1980s and 1990s, when Rathje's views were being widely circulated, concerns about the toxicity of various additives found in plastics were low, and worries about how microplastics could be impacting ecosystems and human health were virtually non-existent. The public's understanding of the extent of plastic in the ocean and the impact this has on sea life and biodiversity was also far lower than it is now.

It was perhaps unsurprising then that Rathje advocated for flexible packages with an eye to reducing waste in landfills, but seemingly gave little thought to any of the other impacts that would prove so problematic in the decades to come.

Today, flexible and multilayer plastics make up about 59 per cent of the plastic packaging produced, but 80 per cent of the plastics that leak into the ocean globally. Around 25,000 flexible packages enter the ocean every second, according to the Ellen MacArthur Foundation. Source reduction has failed to stem the rising tide of plastic waste. In the US, plastic waste per person jumped from about 60 pounds in 1980 to 218 pounds in 2018.

Through the 1990s and early 2000s, the impact of selling billions upon billions of light, cheap, flexible plastics was felt particularly in emerging markets, as the next chapter will show. Such places already suffered from a shortage of organised waste collection and used flexible plastic packaging was widely dumped and littered. While rigid plastics like bottles have enough value to be attractive to waste pickers,[3] used flexibles are so light and low-value as to be worthless.

'I would expect that Bill might feel differently today,' says Lilienfeld, who believes the flexible plastic packaging industry has been a victim of its own success. 'Today, there is so much flexible packaging in use that the original benefit has been forgotten in the sea of plastic bags and pouches that have followed. Rather than take the public concerns seriously, the industry generally tries to hide behind its source reduction benefit.'

---

[3] Waste pickers are informal workers who often trawl through dumps and landfills or pick up street litter to collect recyclables. According to a 2020 report from the PEW Charitable Trusts and SystemIQ, they are responsible for 60 per cent of global plastic recycling.

PART III

SUPERCHARGING CONSUMPTION

CHAPTER 9

MAKING A MARKET OF THE WORLD'S POOREST

'You need to convince consumers that their proxy is not good enough anymore'

In 1990, the average Indian consumed 0.7 kilograms of plastic a year. That compared with 100kg in the US and a world average of 12kg.

Despite the lack of formal recycling programs, the culture in India had long been to reuse or recycle things. Every household that got newspapers kept them in neat piles to sell to *raddiwalas* – scrap collectors – who in turn sold them on for recycling. Used clothes were mended or traded in for stainless steel kitchenware, while old saris were turned into curtains and cushion covers. Ghee tins were used to store spices, while old calendars were used to wrap textbooks. Cooking oil was used more than once, fruit peels were turned into face masks, bones and scraps were fed to pets, livestock or neighbourhood strays, and shoes were handed down.

Outside the home, tea was dispensed in tiny clay pots, freshly fried snacks were served on pieces of newspaper, and rice and lentils were sold in jute bags. Fast-food restaurants selling *dosas* and *idlis* used washable steel plates. In villages, farmers wore sandals fashioned out of used cart tyres.

Door-to-door pushcart vendors sold fresh, unwrapped vegetables every day. Weddings served thousands of guests on banana leaves. Soft drinks came in returnable glass bottles – people milled about to drink them outside shops before handing back the containers. Office workers ate home-cooked meals from steel tiffin boxes – an elaborate network of *dabbawalas* collected and returned the containers to people's homes so that white-collar workers didn't need to tote the multi-compartment *dabbas* to and from work.

But by the late 1980s India's culture was quickly changing.

The best minds at Western consumer goods companies had set their sights on conquering India's sprawling, and largely untapped, market. Nestlé was selling Maggi instant noodles. Unilever had an array of detergents and shampoos. Colgate's toothpaste, toothbrushes and soaps were widely available. Johnson & Johnson was pushing its sanitary pads and baby powder, and Cadbury was selling chocolate and powdered drinks. They all used plastic packaging, joining an array of Indian companies like Godrej and Parle who were also relying on plastic to sell their products up and down the country.

In the summer of 1995, the *Times of India* published a telling interview with a small fruit seller who complained that customers had abandoned him for other sellers who offered plastic bags. People who had long brought their own bags or a clean handkerchief to carry fruit home in now demanded the convenience of a disposable bag provided on the spot. 'If I did not give my customers a bag, they would not buy from me,' he told the newspaper. 'It became a fashion and I also started handing out plastic bags.'

By 1997, India's consumption of plastic stood at 2kg per person – and Mumbai's stood at 6kg. It was still very low compared to the rest of the world, but multiply 2kg by nearly a billion people and you get a lot of plastic. Plastic waste had become rampant enough that India's environment ministry set up a special task force charged with curbing litter and promoting recycling. Video jockeys from popular music video channel, Channel [V], began urging their young viewers to stop using plastic bags.

Many executives at Western multinationals saw the building concerns about waste as having little to do with either them or the billions of plastic-wrapped products they were churning out. Much of what they sold was in small packages, especially single-use sachets. Made from layers of plastic tightly bonded together, the sachet was the ultimate flexible package in terms of both price and functionality, while offering the source reduction benefits that Rathje extolled. It was light, robust, portable and weather-resistant. At 2.5 inches by 1 inch, it could be discreetly tucked into a sari blouse or slipped into a pocket.

The sachet was also the cheapest product on the market, aimed at low-income Indians, many of whom earned daily wages and so didn't have the luxury of buying large quantities of shampoo, oil or detergent. Government officials charged with tackling waste largely ignored it. Instead, they cracked down on more visible culprits like plastic bags, leaving consumer goods companies to keep selling billions of sachets each year, many of which were littered on streets and in waterways.

'The worry, even if it was there, was quite low,' Vivek Bali, P&G's one-time haircare head for India, remembers. 'We do know that many things in India get recycled.' He pauses a moment, before adding: 'But very frankly there was no strategy to get sachets back into a recycling process.'

*

Consumer goods companies in the West had occasionally used sachets to entice consumers to try new shampoos and laundry detergents. But it was in emerging markets like India that companies realised the full potential of the plastic sachet.

As a mainstream, everyday package, the sachet unlocked enormous multi-billion-dollar markets that had long been out of reach, promising a rush of new growth, courtesy of some of the world's poorest people, just as sales in developed markets started to slow.

Indian shopkeepers had long found their own ways to make products like soap and laundry detergent affordable for the hundreds of millions of Indians – particularly in the country's villages – who couldn't afford a full bar or bag. They opened up 1kg bags of washing detergent powder and sold 100 grams at a time for 1 rupee (which is well under 1 cent), dispensed into people's own containers. They cut bars of soap into smaller pieces, selling these individually.

By the late 1980s, Unilever's India subsidiary – Hindustan Lever – had begun packaging its shampoo brands Sunsilk and Clinic in sachets that held just enough for a single wash. 'There was a realisation in Unilever decades ago that people on daily wages also have the same aspiration as people on monthly wages,' says Vindi Banga, former chairman of Hindustan Lever. 'That is the genesis of the single-use pack.'

Unilever had been selling its products in India since 1888, when crates prominently printed with the words 'Made in England' landed at Calcutta's harbour. The crates carried Sunlight soap. More products followed and, by 1931, Unilever had set up its first Indian subsidiary, making *vanaspati*, a low-cost substitute for ghee.

Over the years, Hindustan Lever's in-house army of psychologists and market researchers had worked to more fully understand the motivations fuelling the Indian housewives who were its main customers. They visited the women's homes and brought groups of them together for tea, spending a long time getting them to relax so they'd talk freely. Among the valuable insights this time-consuming method of market research yielded was that the company's relatively pricey soaps were often bought by poor women.

'It is not necessarily the rich woman who buys Sunlight,' explained Hindustan Lever's chairman S.H. Turner at the company's annual shareholder meeting in 1958. 'It may be the poor woman who values its lather and the ease with which she can do her washing with it; the rich woman whose washing is done for her by servants often feels that they are so wasteful to [*sic*] soap anyway that it is best to buy for them the cheapest brand on the market.'

Hindustan Lever's researchers also found that there was little knowledge of the company's brands, particularly in rural India. Farmers spent the most on food, tube wells, fertilisers, seeds and farming equipment. Weddings were a high priority too − these needed to be opulent in order to raise the family's standing. Next came consumer durables like bicycles, sewing machines and radios. At the very bottom were the kind of packaged goods Hindustan Lever sold, like soap and vanaspati.

To convince Indians they needed products they had long done without − like liquid shampoo and powdered laundry detergent − Hindustan Lever deployed its multi-million-dollar marketing machine. 'All consumer goods manufacturers must spend a lot of money on advertising − particularly in India,' said Turner in his 1958 speech. Advertising, he explained, 'may often be required to introduce the product as one which is a necessary part of a higher standard of living'.

Decades later, in 2014, Unilever's south-east Asia head would once again describe the industry's approach to demand creation, outlining for investors how Unilever was creating a market for dish-washing soap. 'You can actually clean dishes with ash and salt. It works,' he said. 'But my task is to convince consumers that there is a better, more hygienic, faster, less residual solution. You need to convince consumers that their proxy is not good enough anymore.'

To create new markets in rural India, Unilever employed a totally different model to the one it relied on in the West. Back in the

1960s, most Indian villages were 'media dark'. Hardly anyone had a TV and most people were illiterate. Hindustan Lever found ways to advertise to them anyway. It had a fleet of 12 vans that drove around villages showing Bollywood and local language films, either on TVs played from the backs of the vans or on roll-down screens. People happily sat through dozens of Hindustan Lever advertisements in exchange for the entertainment.

Another seven vans drove around doing demonstrations: Hindustan Lever's marketers showed people how to use its soaps and detergents by physically washing clothes and half-clothed bodies in public. They performed skits and puppet shows. They travelled around the country painting advertisements for Unilever's brands on the walls of village compounds.

Operating and maintaining the vans across enormous distances on bumpy rural roads was expensive.[1] The cost of marketing to each viewer of Unilever's rural cinemas was 25 paise, a whopping 125 times higher than the 0.2 paise it cost to similarly reach an urban consumer.

The efforts didn't justify the paltry sales of low-cost soaps and cooking oil they drummed up. But Unilever stayed the course. Its polling showed that once workers received regular pay, their interest in buying packaged food and soap rose. Unilever's executives saw the potential to turn millions of Indians into consumers by stirring what the company claimed were their nascent desires.

Key to making money from low-priced (and hence inevitably low-margin) products was driving enormous sales volumes – selling billions upon billions of products. 'Economies come largely from mass production and mass selling of standardised lines,' Turner told Hindustan Lever's shareholders in his 1958 speech. 'We must have products which are acceptable alike to the rich and the poor.'

---

[1] A sales van cost 2.5 times that of a non-mechanised salesman, while the selling cost per unit was 17 times higher.

The single-use multilayer plastic sachet, which Unilever would only successfully roll out some three decades later, enabled the kind of mass selling that Turner knew was needed to turn a fat profit in a poor country. Sachets would become Unilever's primary way to level the playing field in India, offering the same products to rich and poor. In time, they'd catch on widely, with thousands of different consumer goods brands opting to use the tiny plastic packs.

'In India, everything is sold single. You can buy one banana, one cigarette, one egg and – whether it's pickle, hair oil or salt – you can buy it in a sachet form,' says Anand Kripalu, who worked at Unilever for more than two decades before leaving in 2005 to become Cadbury's south Asia head. 'The best example of how sachets exploded a category is shampoo,' he adds. 'They put shampoo within reach of hundreds of millions of people overnight.'

*

The practice of shampooing has a centuries-old history in India. In fact, the word derives from the Hindi *champo*, meaning to press or massage. But among the revolving door of India's finance ministers, for a long time commercial shampoo was seen as an indulgence used by a small slice of the country's upscale urban population. The government taxed shampoo as a luxury product, which of course ensured that's what it remained. Shampoo bottles of 200 millilitres were priced at about 25 rupees (30 cents), which was unaffordable for most people.

And so most Indians ignored commercial shampoo. They washed their hair using homemade concoctions made from *reetha* (soap berries) and *amla* (gooseberries) or just used body soap. They relied on coconut oil to condition it. The few who used packaged shampoo did so for weddings, religious festivals or birthdays.

The man who would first try to convince regular Indians to buy shampoo en masse wasn't a Unilever executive but rather a one-time schoolteacher called Chinni Krishnan who hailed from the port city of Cuddalore in Tamil Nadu. In the late 1960s, he quit teaching to start a business repackaging large bags of powdered pharmaceutical products into small portion sizes that poor people could afford.

Making single-use sachets for dry products was easy if labour-intensive. India's loose-leaf tea sellers had for years been selling small portions of tea in paper packets for as little as 3 paise. Chewing tobacco makers employed similar sachet packaging, made from thin metallised plastic.

But it was Chinni Krishnan who figured out how to make plastic sachets that could hold liquids. In 1972, he cut a hosepipe down, sealing both its ends using a machine designed to seal plastic file folders. He later shifted to using softer polyvinyl chloride tubes, creating puffy sachets that resembled tiny plastic pillows. Chinni Krishnan tried selling shampoo in his new sachets, creating a brand called Velvette, but in the absence of marketing it didn't sell particularly well.

His efforts caught the attention of Hindustan Lever. The consumer goods giant briefly tried its hand at selling PVC shampoo sachets in the mid-1970s, but pulled the plug in less than a year.

Shopkeepers found the transparent, pillow-shaped pouches difficult to store; their shape meant they couldn't be stacked and had to be kept in bins. Consumers found the sachets hard to open, particularly with wet hands. For Unilever, they were expensive to make and sell. 'The margins were half of what the bottles were,' recalls Nihal Kaviratne, who worked as Hindustan Lever's marketing manager for personal care in the early 1970s. 'Although the unit price was low, on a per-wash basis they were also very expensive for consumers.'

In time, Chinni Krishan's son, C.K. Rajkumar – who took over the business after his father died – found a way to make the Velvette sachets easier to open. Another son, C.K. Ranganathan, launched his own shampoo brand in sachets, Chik.

Just as Hindustan Lever's salesmen had shown villagers how to use bar soap, Ranganathan's new band of employees found models – usually schoolboys easily swayed by the promise of free sweets – whose hair they could publicly shampoo as the rest of the village gathered around and gawped. 'We applied shampoo on one half of the head and on the other we applied soap – the split head method,' says Ranganathan. 'After washing, and once towelled and nicely dried up, we asked the public to touch and smell the soap part and the shampoo part.' Chik's sales began to tick up.

Despite its desire to sell to rural Indians, Hindustan Lever remained focused on pushing products like vanaspati, bar soap and fertilisers. 'The definite policy at Hindustan Lever at the time was not to be seen as a company selling luxury items,' says Alec Lever, a voluble and floppy-haired Englishman who began working at Unilever's India business in 1985. 'They didn't want to be seen as a multinational making products for a westernised elite.'

Lever, no known relation to the company's founder, had arrived in India from the Philippines, where P&G was using sachet packaging for its Prell shampoo. Prell was sold in the form of a solid green needle that foamed when mixed with water. Lever copied the sachets, but imported Hassia packing machines from Germany to fill them with liquid Sunsilk. The liquid shampoo was far easier to work into hair and made for a more luxurious experience than needles. Sunsilk's sales shot up and liquid shampoo in sachets quickly became the norm across the Philippines.

India's growth had long been so sluggish that economists despairing of its plodding socialist polices nicknamed it the 'Hindu' growth rate. But by the late 1980s, when Lever arrived in India, this was changing. Crop yields were rising, jobs were opening up in rural India courtesy of manufacturers lured by government tax holiday schemes, and better roads and more television ownership were making it easier for brand-owners to reach villagers.

Hindustan Lever's board gave Lever the green light to try to convince the hundreds of millions of Indians who had never given shampoo a second thought to buy Sunsilk and Clinic. While Chik and Velvette were already being sold in South India, Hindustan Lever had enormous national reach. Suddenly, shampoo sachets started showing up across the country.

The new versions were made from several layers of two types of polyethylene and aluminium, and were much easier to handle than the old PVC ones. A strange quirk about the new sachets was that, on a per ounce basis, they were priced cheaper than larger bottles, flying in the face of long-established industry precedent that bigger is cheaper. One theory is that demand for flimsier plastic sachets from the smokeless tobacco or *gutkha* makers dotting every street corner created a big sachet market long before the package was used for shampoo, driving packaging prices down. Another is that, given how large and promising the Indian market was, companies were willing to subvert precedent if it meant building demand.

For women, who typically were in charge of buying house-hold essentials, the appeal of sachets went beyond price. They offered built-in portion control – precluding wayward children or careless husbands from using too much, the way they might from a bottle.

Less than a quarter of villagers had access to running water at home. Many bathed by a local river or water body, or used buckets of water drawn from the village well. This made sachets – designed to be used once and thrown away on the spot – far more convenient than bottles.

Shopkeepers and distributors liked the new sachets too. They held up well against humidity and were perfect for India's extreme heat and bumpy roads. They didn't take up valuable counter space at the tiny *kirana* shops that most Indians frequented since they didn't need to be stacked. Instead, they hung on strings and were light-reflective, acting as colourful advertisements for Sunsilk and Clinic Plus, even in badly lit markets that often operated in candlelight amid India's notoriously frequent power cuts. 'There weren't many things being hung from shop ceilings, so we had to have someone with a hammer and a nail going around,' recalls Harsh Bahadur, who worked closely with Alec Lever as a product manager for Hindustan Lever's Clinic and Sunsilk shampoos.

Bahadur in the late 1980s commissioned a study to figure out how Indians washed their hair and how they viewed shampoo. It revealed that the average person thought they needed to empty a shampoo sachet into a mug of water, mix it and tip it over their head. Unsurprisingly, this over-diluted the shampoo, rendering it ineffective and leaving people disappointed. 'Shampoo had 1 per cent or 2 per cent penetration[2] and we figured from that study that there was a complete lack of information about how you use it,' says Bahadur. 'People didn't use shampoo. People didn't know what shampoo was.'

Unilever began running TV ads to demonstrate how to use shampoo. Notched, easy-to-tear sachets weren't yet a thing and so an actress demonstrated how to cut open a sachet using a pair of scissors, squeeze the shampoo out and massage it into wet hair. 'The whole idea was to get hundreds of thousands and then millions of people to use it,' Bahadur says.

---

[2] Penetration is an industry term for the percentage of households who use a product.

The new exciting variants of Sunsilk come in
bottles of appealing design as well as handy
sachets.

A July 1987 ad for Sunsilk shampoo shown in Hindustan Lever's
in-house magazine. Image courtesy of Unilever

Given shampoo's niche appeal, Hindustan Lever's ad agencies hadn't bothered to create ads tailored to local sensibilities, instead repurposing Unilever's Western campaigns, which were totally out of sync with how the vast majority of people lived. A 1980s advertisement for Sunsilk shampoo showed a young woman with flowing shiny hair jogging down Bombay's upmarket Marine Drive promenade in shorts, showing enough leg to make the average sari-clad Indian woman avert her eyes.

'While Sunsilk was being advertised to Sandra in Bandra, our target market was Geeta in Gorakhpur,' says Alec Lever. (Bandra is an upscale part of India's Bollywood capital, Mumbai, and Gorakhpur a relatively overlooked city in the northern state of

Uttar Pradesh.) The personal care team did away with women in shorts, as well as with overt depictions of romantic relationships that made many Indians uncomfortable. Instead, it showed sari-clad wives and groups of friends.

Slowly, the combination of sachets and Unilever's new advertising had the intended effect, including in rural India. By 1989, 6.5 per cent of India's shampoo was being bought by people in rural areas, up from close to zero only a few years earlier. To further encourage sales, Unilever offered women one free hair pin with every Sunsilk sachet they bought.

Then, in August 1990, the Gulf War hit, sparking an economic crisis in India. Import bills soared, exports plunged and foreign exchange reserves dwindled as remittances from Indians in the Middle East evaporated. In exchange for a loan from the International Monetary Fund and the World Bank, India agreed to liberalise its economy. As part of this, in 1993 India's turbaned finance minister Manmohan Singh slashed excise duties on shampoo, along with other products like talcum powder and face cream.

The tax cut was just one of many changes that would make India far more attractive to multinational companies long put off by its protectionist, state-controlled economy. 'India was opening up and every man and his dog wanted to get in,' remembers Banga, the one-time Hindustan Lever chairman.

By the early 1990s, firms no longer needed to lobby Delhi's bureaucrats to get industrial licences to set up new facilities or expand existing ones. No longer was foreign ownership capped at 40 per cent; companies could now own 51 per cent as a matter of course, or even beyond if they applied for special approval.

And no longer was shampoo treated as a luxury product. Taxes dropped from 120.75 per cent to 70 per cent, making sachets way more profitable. Almost overnight, there was much more money for Unilever and other shampoo makers to spend on advertising and distribution. Over the next few years, duties continued to fall and, by 1997, they stood at 30 per cent.

'Suddenly the population woke up to shampoo,' says Sanjiv Kakkar, who worked as marketing head for Hindustan Lever's personal care business for five years from 1993 onwards. 'And they had easy access to it because it was in sachets.'

*

Liberalisation also unleashed another big force that would super-charge shampoo sales. It came in the form of Unilever's arch nemesis: Procter & Gamble.

The fierce global rivalry between the two companies had played out for decades. As early as 1930, Unilever executives were bristling at P&G's encroachment onto their turf after the Cincinnati company acquired a soap manufacturing firm in north-east England. Since then, the pair had tussled over soap, shampoo and detergent in the US, Europe and Asia.

Before India's liberalisation, P&G didn't sell shampoo in India. When the economy opened up, it saw an opportunity to jump in. Even compared to other emerging markets, the runway for growth was huge: China and Indonesia's per-capita shampoo consumption respectively stood at more than seven times and eight times that of India. 'India was one of our future focus markets,' says Rahul Kadyan, who was hired by P&G as an assistant shampoo brand manager in 1994 at the age of 22. 'It was growing rapidly and the middle class was coming into its own. This was a big bet for P&G.'

One of P&G's first products in India was Ariel, an enzyme-based detergent powder targeting the wealthy Indian women marketers called the 'silk sari brigade'. Hindustan Lever had rushed out a rival detergent, Surf Ultra, before P&G could get Ariel on shelves. Then, a nasty rumour about Ariel swept kitty parties.

Women at the gossipy social gatherings popular in Delhi, Mumbai and Calcutta confirmed they had heard that washing with the new P&G detergent faded clothes.

P&G countered by getting a major textile manufacturer to endorse Ariel. Unilever complained to regulators about the endorsement, eventually forcing P&G to change its ads. 'We were determined not to allow P&G to have a successful beachhead in India and therefore we made it our absolute mission to block and destroy them,' says Vindi Banga, who at the time was marketing manager of Hindustan Lever's laundry detergents business.

Faced with Unilever's offensive, Ariel had turned into a far more expensive launch than P&G had budgeted for. To take on Hindustan Lever's shampoo business, P&G knew it needed to pull out all the stops. 'Basically, India was almost a Unilever monopoly,' says Kadyan. 'There were a couple of other players locally, but nationally it was really Unilever which owned the shampoo market. The whole idea was "how do you unseat a behemoth?"'

In July 1994, during his induction at P&G in Mumbai, Kadyan was handed a copy of the ancient Chinese military treatise, *The Art of War*. The book was a favourite of P&G's India CEO, who impressed upon his new recruits the importance of the tactics it espoused. 'All warfare is based on deception,' it read. 'Hence, when we are able to attack, we must seem unable; when using our forces, we must appear inactive; when we are near, we must make the enemy believe we are far away; when far away, we must make him believe we are near.'

Later that day, Kadyan learned he'd be working on a top-secret launch for a brand whose name was never to pass his lips, not to his parents, not to his friends, not even to fellow P&G staffers. Kadyan, like the rest of the launch team, was to refer to it only as 'Project Gleam'. Gleam was code for Pantene, the biggest shampoo brand in the world and a huge money-maker for P&G.

Many of Kadyan's university friends from the Indian Institute of Management in Lucknow had landed jobs in Mumbai, and the group's drinking sessions at Cafe Mondegar – one of the city's oldest and most characterful bars – invariably slid into bouts of boasting. Kadyan longed to join in, but took his vow of silence seriously. 'Everyone knew I was working on a top-secret project, but nobody knew what it was,' he says. 'It was all a bit Sherlock Holmes.'

In advance of Pantene's launch, P&G booked airtime with major TV channels. Anticipating that once Unilever realised it was launching a new brand the company would buy up all available space, P&G told media executives that the slots were to advertise Ariel. Still, it didn't take long for Unilever's sources to funnel back to the company that P&G was launching a shampoo brand in India.

Kadyan's boss at P&G, Vivek Bali, decided the best way to throw Unilever off the scent would be to pretend the brand was Head & Shoulders. He instructed market researchers to canvass consumers for their opinions about the anti-dandruff shampoo in areas they knew Hindustan Lever's employees lived. P&G's creative agency, Grey, printed small batches of prototype labels for various shampoo brands, carelessly 'forgetting' these in printing machines across Mumbai.

'They overspent their ad budget on all these me-too ads for the different brands that they thought we might launch,' remembers Udaiyan Jatar, a Grey executive who worked on Pantene's India launch. 'Of course, the word Pantene never escaped our lips. It was very cat and mouse.'

Pantene's ads were nothing like those Indians had seen to date. They featured fantastically stylised shots. 'There was the ribbon shot where you'd find a way for the hair to unfurl by pulling on a ribbon or silk scarf, and there was the waterfall shot where you'd lift the hair up with one hand and then pull it away and it would all tumble down,' explains Kadyan. Using two hands to lift the hair was dubbed the 'peacock' shot since it resembled a fanned-out tail.

Pantene's launch coincided with the transformation of India's cable industry. During the Gulf War, dish antennas had sprung up across rooftops in major cities to deliver cable news from the US. Satellite TV quickly took off and racy new shows in Hindi and local languages were shown alongside American soaps like *The Bold and the Beautiful* and *Santa Barbara*. The glamorous women on screen wore their shiny hair long and loose, the opposite of the tightly bound, heavily oiled buns and braids most Indian women had long favoured. 'There was a cultural shift,' says Jatar. 'It was serendipitous and our timing turned out to be good.'

Despite Pantene's global heft, most Indians had never heard of the brand. P&G's internal research showed that washing once with Pantene made hair feel shinier and softer. Vivek Bali, P&G's India haircare head, decided to blitz households with millions of free 6ml sachets of Pantene, betting that a single sample would be enough to win new consumers.

Starting in July 1995, through the four months leading up to Pantene's launch, an army of promoters coached by Kadyan knocked on the doors of 13 million households across 23 major cities in India. They travelled on foot in the blazing heat. About half the samples they handed out were to people who didn't use any shampoo at all.

But P&G didn't just use sachets as samples. When it launched Pantene in November 1995 across India's biggest cities, it sold the shampoo in both a plastic bottle and a sachet. 'At that time, there was a very high contribution from sachets and so we were riding on a habit people already had,' explains Bali.

Sales took off like a rocket. In just five months, Pantene captured more than 10 per cent of the shampoo market. It was steadily gaining ground on Unilever's brands Sunsilk, which had 17 per cent of the market, and Clinic, which had 25 per cent.

P&G's India head, David Thomas, told investors he was confident that Pantene could win 20 per cent of the Indian shampoo market. 'It was the fastest growth of any launch Pantene had seen anywhere in the world,' says Kadyan. 'A lot of it had to do with the scale and speed of the sampling programme which, by the way, was enabled by the sachet.'

Just as the rivalry between Kimberly-Clark and P&G had turbocharged America's diaper market, Pantene's success sparked a massive scramble for share among rivals, which ultimately resulted in a huge boom in India's shampoo market. In 1996 alone, Indian shoppers were introduced to 132 new shampoo brands and variants. Between 1993 and 1996, the amount of shampoo Indians purchased annually climbed 89 per cent, while the number of households using shampoo jumped 50 per cent.

To fight Pantene, Sanjiv Kakkar of Hindustan Lever sent sachets of a new 'root-nourishing' shampoo called Organics to 10 million houses in Delhi, Mumbai, Calcutta and other cities. 'When my hair breaks, my heart breaks' was the melodramatic tagline.

Unilever also redoubled its advertising efforts across its other shampoo brands. 'It became a competitive slugfest,' says Kakkar. 'Suddenly shampoo became one of the most heavily advertised categories and you couldn't switch on the TV without seeing one of our ads.'

*

By the late 1990s, a common statistic bandied about the conference rooms at Hindustan Lever's Mumbai offices was that India was home to 16 per cent of the world's population but had 28 per cent of its hair.

Despite this – and the rapid growth in shampoo sales over the past decade – India accounted for not even 3 per cent of the world's shampoo market. A large barrier holding people back was price. By now, the majority of shampoo sales in India were made using single-use sachets. But even at a few rupees per wash, shampoo remained beyond the reach of much of the country. This was particularly the case in rural areas, where some 40 per cent of people were paid on a daily basis.

C.K. Ranganathan, the founder of Chik shampoo, had long priced his 6.5ml sachets of Chik shampoo at 1 rupee, which was half the price Unilever charged for its sachets. Then, in 1997, while Unilever was still preoccupied with fighting P&G, Ranganathan noticed that shoppers had begun asking for cheaper products. Mobile phone usage was exploding and many low-income consumers used the extra savings to buy a daily phone recharge, yet another single-serve innovation that was sweeping India.

Ranganathan decided the time was ripe to lower the price of sachets. He asked his suppliers to concentrate Chik's formulation. Since shampoo is mixed with water anyway, his reasoning was that this wouldn't hurt Chik's efficacy. In 1998, he launched the more concentrated dose in a 4ml sachet for just 50 paise, half the price of his original offering. The price cut effectively upended the market and accelerated India's shift to the tiny plastic pouches. Overnight, shampoo had become affordable to tens of millions more people, especially in rural areas.

Hindustan Lever was reluctant to follow suit, since slashing prices invariably cut into margins. By now, its Sunsilk and Clinic sachets cost Rs 2.50. That put the opening price for both shampoos at five times that of Chik. Instead, Hindustan Lever launched a marketing campaign. It promised to give Indians who bought Clinic Plus and Sunsilk the chance to win 2.5kg of gold. There was an initial rush to buy the sachets, but it inevitably tailed off.

Shiv Shivakumar, who took over from Kakkar as Hindustan Lever's shampoo marketing head in 1998, was convinced the company needed to slash prices. For months he badgered the board to let him follow Ranganthan's lead and launch 50 paise sachets of Sunsilk and

Clinic Plus. His pleas fell on deaf ears. One senior executive 'said CP stands for Clinic Plus not Consumer Promotion,' recalls Shivakumar, an energetic man with a penchant for Bollywood film songs.

It was only in August 2002, as Chik was on the verge of vaulting over Clinic Plus to become India's largest shampoo brand, that Hindustan Lever's board capitulated. Shivakumar was allowed to slash the price on Sunsilk and Clinic's existing sachets from Rs 2.50 to 2 rupees and also launch new, smaller 50 paise versions.

By 2002, shampoo penetration had jumped to 46 per cent of households, up from 18 per cent in 1995. Urban penetration had doubled, but it was really rural households driving the shift, with penetration rising three-and-a-half times. It was a massive jump. Villages now accounted for 40 per cent of shampoo sales. But Unilever knew there was room to go even further.

*

In April 2000, Keki Dadiseth, Hindustan Lever's portly, bespectacled chairman, announced that the company was looking at new ways to get its products into the furthest reaches of the country. The rural push, Dadiseth told Hindustan Lever's shareholders, was key to ensuring that Hindustan Lever achieved its ambition to double its sales every four years. 'It is our view that India will soon see an inflection point in rural consumption,' Dadiseth – who was in his last few months as chairman before Vindi Banga took over – told the shareholder meeting. 'Rural demand and consumption of consumer products is set to explode.'

Hindustan Lever's sprawling distribution network across rural India had long been one of its biggest strengths. But competition had intensified and this advantage was fading. Hindustan Lever's distributors reached only 11 per cent of India's 627,000 villages.

The remaining villages, home to populations of fewer than 2,000 people, had unpaved roads reachable only by bullock cart or bicycle. The company had dismissed them as 'unviable and inaccessible'.

There was a high cost to sending its own people out to far-flung locations the way Hindustan Lever had in the days of roving movie vans and live demonstrations. Yet the company was reawakening to the idea that selling large amounts of low-margin products could, in time, make for big profits. After all, India's rural residents made up 12 per cent of the world's population.

There was also an opportunity cost to staying away, providing an opening for competitors like low-cost washing powder Nirma, which had grabbed large amounts of Hindustan Lever's market share, and, of course, Chik shampoo. As India opened up and P&G entered, Chik's parent company CavinKare had benefited from the advertising storm that followed. Many villagers simply asked for 'shampoo' when they turned up at a store, and they wanted it on credit. Chik was usually the cheapest option and Ranganathan offered a higher profit margin to shopkeepers, incentivising them to push his brand.

'The challenge for most companies is to be able to offer appropriate products in an affordable way in relatively remote locations,' said Dadiseth at the 2000 meeting. 'We are now convinced that the answer to this challenge lies in setting up a specific rural business system.'

That system turned out to be tens of thousands of India's poorest women.

Hindustan Lever began running massive recruitment drives in tiny villages, signing up women whose average household incomes were less than 1,000 rupees a month to become its distributors. Unilever called them 'Shakti Ammas': *shakti* means strength in Hindi and *amma* means mother. It arranged loans for the women, loading them up with a minimum of 10,000 rupees in shampoo, soap, detergent, skin creams and other products each month. Think of it as the Avon model, but for India's poorest.

It was a gamble. Most of the women were housewives with zero sales experience. Selling to their neighbours, friends and local store owners required chutzpah, and not everyone who applied to be a Shakti Amma was accepted. 'There had to be that element of leadership amongst the women. They had to be more confident and outgoing than the rest,' says Sharat Dhall, who was appointed to head up Project Shakti soon after it began. Caste was also a consideration. 'The low-caste women were too scared to go out, the high-caste ones didn't want to. It had to be a middle caste so they weren't shunned by either side.'

To help its new distributors build confidence, Unilever hired what it called rural sales promoters, essentially cheerleaders who worked six days a week, visiting up to five villages a day to give advice and pep talks to the Shakti Ammas.

Project Shakti – which today counts 200,000 Shakti Ammas as Unilever emissaries – helped Unilever increase sales of its shampoo, soap and laundry detergent across India. That included selling sachets in the country's most remote areas, places with no organised waste collection where much of what was discarded had long been organic.

C.K. Prahalad, a University of Michigan management professor who was also a member of Hindustan Lever's board, took a great interest in Project Shakti. Influenced in part by what Unilever and others were doing to sell their products to people on low incomes, Prahalad went on to write an article entitled 'The Fortune at the Bottom of the Pyramid', followed by a popular book of the same name. He made the case for how other companies should tap into an economy he estimated was worth $13 trillion worldwide by catering to the billions of poor consumers they had largely ignored.

Prahalad's views attracted some criticism. 'Professor Prahalad is correct when he says everyone wants quality services and products, but which ones?' wrote Atul Wad, a former director of the international business development programme at Northwestern University, in 2004. 'Selling shampoo in rural India is great for the producers, but with the very limited household income most people have, is this the best way for them to spend their money?'

It's easy to interpret Wad's comments as offensive. At first blush, they seem to imply that people on limited household incomes shouldn't be free to choose how to spend their own money. But assuming that anyone is choosing freely ignores how much of the 'market making' companies undertake relies on seeding or deepening insecurity about our existing way of life, thwarting our ability to make choices that truly are best for ourselves.

For decades, Unilever ran TV ads for its skin-lightening cream, called Fair & Lovely, implying that young women who didn't have fair skin wouldn't find jobs or husbands. Fair & Lovely went on to become Unilever's largest personal care brand in India.[3] Today in India, Unilever's antiperspirant brand Rexona organises 'confidence building seminars' among young women in university. The seminars underscore the importance of preventing body odour for any young woman who wants to land a job or be successful. Before the Rexona Confidence Academy, the idea of using deodorant had never been considered by many who instead neutralised odour naturally by wearing jasmine in their hair, washing their clothes and bathing.

Everyday examples of market making include convincing men that they need male grooming products and persuading women using face cream that they need more expensive face cream, eye cream, wrinkle-reducing cream and hair-removal cream. They include convincing people who cook for their pets or feed them

---

[3] In 2020, in the wake of the Black Lives Matter movement, Unilever renamed the cream Glow & Lovely after consumers and non-profits accused it of building upon, perpetuating and benefiting from racism.

table scraps that they need packaged pet food, and new mums that they need disposable diapers. They include telling anyone using shampoo that they need conditioner too, and anyone using conditioner that they also need hair serum and leave-in conditioner.

Today, social media makes this far easier than ever before, as companies, often insidiously, harness influencers to convince people to buy more stuff, 'better' stuff – all in the name of upgrading their lifestyles.

Back in the early 2000s when Unilever launched Project Shakti, some rural development non-profits complained that the company was exploiting poor women in India – and putting their families at risk – by encouraging them to take on debt.[4] Unilever batted away such criticism, comparing its efforts to the micro-finance pioneer, the Grameen Bank. 'For people in rural India, this translates into critically needed, sustainable jobs contributing to better living standards and prosperity,' said Dadiseth. 'For us, it translates into access to hitherto unexplored territory.'

*

In 2001, the *Lucknow Times* published an explosive article quoting a cow shelter owner who said 100 cows a day were dying from eating plastic bags in the north Indian city of Lucknow alone. 'The affected animal will have a skeletal body but abnormally bloated stomach. It will very eagerly wobble to the trough but would only sniff at the fodder, unable to eat anything,' wrote

---

[4] In 2011, a study of 15 current and former Shakti Ammas in the state of Madhya Pradesh by University of Guelph associate professor Marta Rohatynskyj concluded that Project Shakti 'does not empower poor, rural women'. Among other issues, it found that much of the time men were acting as dealers and that the poorest women had abandoned the project.

the reporter. 'They gradually become weak due to starvation and then finally become immobile.'

The sight of cows munching on dumped plastic in both cities and villages was far from unusual. The post-liberalisation boom in consumer goods had translated into a sea of plastic waste. Over a few years, India was deluged by bags, bottles, cutlery, cups, diapers, sanitary pads and, of course, sachets.

India, like other parts of Asia, was also taking huge volumes of allegedly recyclable waste from the US and other developed countries. The plastic bottles that could be extracted for recycling were removed. Much of what remained was dumped or burned. Between 1990 and 1993, India imported 19 million kilograms of plastic waste, all of which the exporting countries smugly logged as having been 'recycled'.

The growing mountains of used plastic were often stored in disorganised markets. Every so often, fires erupted, releasing toxic fumes. During the torrential monsoon season, plastic bags regularly clogged drains in big cities, causing floods that washed into low-lying homes and brought traffic to a standstill. In mountainous areas, the blockages triggered landslides, propelling an expert committee to advise that plastics be banned in such areas.

The cows' plight got people talking. For the first time, executives began to worry about a broad crackdown on plastic waste that could include sachets. 'Knowing the importance of the cow in Indian culture, we knew this could at some stage become a hot potato for us to handle,' says M.K. Sharma, who worked as vice chairman and general counsel of Hindustan Lever. Sharma and Shiv Shivakumar flew to Singapore to meet regional Unilever executives to discuss ways to cut down on sachet waste.

Years earlier, Unilever had tried to sell products unpackaged, asking shopkeepers to dole out small quantities of margarine and laundry detergent from larger barrels into shoppers' reusable containers. The effort was short-lived. Nihal Kaviratne, who worked as Unilever's Indonesia chairman in the late 1990s and early 2000s, says eliminating the packaging commodified Unilever's products and erased its most valuable assets: its brands.

Yet, by the early 2000s, executives from Unilever's businesses across Asia were starting to worry that plastic waste could trigger new restrictions, hurting sales. 'This stuff, it began to show up in landfills and nature with our brands on it and that was ghastly,' says Kaviratne. 'We were embarrassed because it could be traced right back to us.'

At the Singapore meeting, Unilever's executives considered incentivising consumers to bring empty sachets back to stores. Back in 1988, Chik had made a splash when Ranganathan began offering a free sachet of shampoo to anyone who brought back five empty sachets. But, more than a decade later, Hindustan Lever decided shopkeepers would be reluctant to do anything similar, given the mess and hassle involved in keeping millions of old sticky sachets in their small stores. 'The cost was not commensurate with the benefits we would derive,' concluded Sharma. Plus, it was unlikely many consumers, even if incentivised, would bring sachets back. 'People washed their hair in ponds, wells or running streams and for someone to bring back a 2-inch by 1-inch sachet was seen as a virtually impossible task,' he says.

Despite India's poor waste collection infrastructure, recycling rates for items like plastic bottles and newspapers were high. Thanks to the country's army of waste pickers, items that had any resale value were quickly snapped up. The shampoo sachets, which were widely littered, had none. They were too small and complex to recycle, so waste pickers ignored them.

Sharma suggested that Unilever, over time, move to paper sachets which, even if littered, would break down. But tests Unilever did on paper revealed major flaws. For one thing, paper didn't hold up against India's high temperatures and humidity. For another, the colourful, high-lustre sachets strung up at storefronts acted as advertisements for Unilever's brands. By contrast, paper sachets were dull, hard to print on and 25 per cent more expensive.

By 2004, sachets accounted for 90 per cent of shampoo sales in rural India and 70 per cent of sales in cities. The tiny plastic packets had also metastasised well beyond shampoo, sparking what the industry's executives proudly called a 'sachet revolution'. Sachets were being used to sell deodorant, toothpaste, face cream, jam, pickle, perfume, ketchup, hair oil, hair dye, pain-relieving balm, hair removal cream, powdered drinks, butter, mosquito repellent, shaving lotion and digestive pills, among other products.

Chik's owner, CavinKare, created a designated 'sachet sales force' to sell the growing number of personal care products it packaged in sachets to tiny shops that had typically sold just cigarettes and *paan* – a popular treat consisting of a betel-nut leaf wrapped around a blend of areca nuts, tobacco and spices. 'Now every product in every category is available in sachets,' trilled a voiceover for a CavinKare promotional video.

Sachets – as with so many other convenience-led products – had spiralled well beyond what its corporate parents had originally imagined. Just as plastic trash bags were once intended to be used mainly by Canadian hospitals, and disposable diapers were dreamed up for weekends away at grandma's, the sachets were conceived of as a way to introduce low-income people to shampoo brands. The idea was that, as incomes climbed, consumers would naturally shift to using bottles. But similar to how plastic trash bags were adopted by everyone and disposable diapers became everyday fixtures for children, the sachets were embraced by Indians of all income levels in both rural and urban areas.

Wealthier Indians liked sachets because they could easily hop between different brands – something companies were always incentivising through marketing that underscored the next best thing. The sachets also fit with the Indian habit of buying small quantities regularly, which was true for those in big cities as well as in villages. For rich and poor alike, there was no financial incentive to buy bottles. The sachets attracted lower sales taxes and it was cheaper to just buy a string of them than one plastic bottle containing the same amount of shampoo.

Kaviratne and Banga are among the many Unilever executives who acknowledge the environmental costs of sachets but justify these by weighing them against the perceived social benefit – that sachets make brands affordable to poor people. The flip side, of course, is that poor people are the ones most impacted by the negative externalities of plastic. Studies show that plastic production and disposal facilities – including dumpsites – are invariably located near low-income communities. At the time, few were talking about microplastics, but we know now that littered sachets break down into tiny plastic particles that spread widely through ecosystems and end up in both animal and human bodies.

In the years that followed, Unilever made other attempts to find solutions to sachet waste across the globe. It encouraged consumers to bring their own containers to refill machines that dispensed detergent, and invested in a door-to-door electric tricycle in Chile that brought the machines to people's homes. The refill projects never scaled. The company ran a trial in Indonesia to recycle sachets through a solvent-based process that separates out polyethylene, the main plastic used to make the sachets. The trial showed that recycling the multilayer sachets was *technically* possible. But just as McDonald's and P&G had found when they tried to recycle foam containers and diapers, the costs of collecting the used sachets were far higher than any revenue derived from selling the recycled plastic, which made the operation economically unfeasible.

'We knew that sachets are the cheapest way to take products to the consumer, but also the most polluting way,' reflects Shivakumar. 'We started looking at solutions in 2001 and until now there is no answer.'

Ultimately, after years of failing to find a solution, Unilever pivoted to publicising its efforts to collect bottles, milk pouches and other 'more viable' plastic waste. On a 2019 reporting trip for the *Wall Street Journal* in which I spent time with waste pickers and walked the streets of Bangalore with Unilever's India executives, I learned that the multilayer plastic Unilever was paying to have collected in order to meet regulatory requirements consisted mainly of larger packets for potato chips and biscuits rather than sachets. Much of what was collected was trucked long distances to be burned at cement kilns, a practice activists decry as dangerous and environmentally unsound. Unilever has since told me that it is no longer 'directly' sending plastic waste to cement kilns. It says it complies with India's regulations and that collected waste is managed in accordance with local guidelines. It didn't address my questions about how sachets specifically are collected or what happens to these. In 2021, nearly 41 billion shampoo packages were sold in India. Of these, 99 per cent were sachets.

CHAPTER 10

FAREWELL TO THE AGE OF ECOLOGY

'We were really training consumers at that time to drink more and more'

On the morning of 30th August, 1999, readers turning to the op-ed section of the *New York Times* were confronted with a smiling portrait of Coca-Cola's CEO, Doug Ivester. His dark hair was slicked into a neat side-parting and he wore his customary suit and tie. He was pictured across from a sad-looking nine-year-old girl clutching a plastic bottle. Beneath was a question from her. 'Dear Mr Ivester: Why do you hurt the Earth when it's so easy not to?'

The provocation was an advertisement from a Georgia-based non-profit called the GrassRoots Recycling Network. GRRN had, for more than two years, been trying to get Coca-Cola to honour a promise it made nearly a decade earlier. In December 1990, a little more than a month after McDonald's said it would abandon the clamshell, Coke and Pepsi had made major announcements aimed at quelling the building backlash against plastic. Within hours of each other, both companies said they would make their soda bottles out of recycled plastic, a first for the industry.

'We needed to do something,' recalls Deborah Cross, who was one of three people who comprised the environmental affairs team at Coca-Cola in the early 1990s. 'It was clear that people were beginning to feel guilty about purchasing the product in plastic.'

If the multilayer plastic sachet was incredibly difficult to recycle, the clear PET drinks bottle was its antithesis. Technologically, it was the easiest plastic package to shred, wash, melt and turn into new pellets.

Environmentally, recycling PET bottles to make new ones was also a win. So far, plastic bottles had been downcycled – recycled into lower-quality products like carpet or strapping that couldn't be recycled further. That meant new plastic bottles had to be made from the by-products of fossil fuels. Bottle-to-bottle recycling, as the drinks giants were now promising, would 'close the loop', reducing waste and greenhouse gas emissions.

In March 1991, in Charlotte, North Carolina, Coca-Cola rolled out 2-litre plastic bottles carrying the label 'Bottle made with at least 25 percent recycled plastic'. Its move had a domino effect. The Council for Solid Waste Solutions threw a splashy press conference, headlined by top executives from P&G and DuPont. They announced that the plastics industry would recycle 25 per cent of all plastic containers – not just drinks bottles – by 1995. At the time, the recycling rate for these was 9 per cent.

Coca-Cola began selling its recycled plastic bottle in 14 other states and in DC, with a large chasing arrows symbol prominently slapped on the front label. It said it would keep rolling the bottles out to more states and handed out bottles containing recycled plastic to every member of Congress.

Within a couple of years, though, Coke and Pepsi – very quietly – stopped using recycled plastic. The plastics industry abandoned its recycling goal for all plastic containers too. By then, oil prices had dropped, making recycled plastic 10 per cent more expensive than using virgin plastic. Used bottles were also in short supply, in part due to America's reliance on exporting its plastic waste to Asia rather than building its own reprocessing plants.

Coke's research showed that consumers wouldn't pay more for a recycled plastic bottle. 'The consumer is a hypocrite,' says Cross. 'They want us to do all the work but they won't chip in 10 cents.' The company's surveys showed it was receiving little acclaim from customers for using recycled plastic. Americans cared about whether a bottle was recyclable. But whether it was made from recycled or virgin plastic didn't seem to register.

Assuaging the public's concern about plastic waste by making bottles from recycled plastic was intended to be a marketing campaign. Given the costs outweighed the benefits, Coke had decided it was no longer in its best interests to continue. 'That's where it went from "This is not a marketing opportunity because marketing opportunities lead to money" to "This is a PR matter",' says Cross.

By now, Jean Statler and Richard Wirthlin were working to develop the 'Plastics Make it Possible' campaign. Coke's strategy was to step back and ride on the coattails of the industry's work to burnish the overall reputation of plastic.

A few years rolled by. Then, in March 1997, Lance King, campaign coordinator for GRRN, fired the first salvo at Coke. He wrote a letter asking the company to make good on its 1990 pledge to make bottles from recycled plastic, and to disclose how much recycled plastic was in each bottle. GRRN was betting that having a consistent and enormous buyer of recycled plastic in place would incentivise further collection, sorting and reprocessing of plastic bottles.

Growing up in Santa Cruz, California, King had his first taste of courting the press as a youth county commissioner in 1969 while still in high school. The job entailed agitating for change on issues of concern to the area's youngsters. If King showed up to school in a suit and tie, teachers knew he had a meeting or rally to attend, and allowed him to skip class. 'We didn't want to be walking around on mousy things like curfews and hostel hours,' says King. At the time, he set his sights on tackling water pollution in Monterey Bay.

King's involvement in GRRN had started in 1996. By then an independent environmental consultant, his personal finances were in dire straits. He had burned through his life's savings after an unfortunate few months trading stocks on margin. He'd also bungled a brief run as an entrepreneur importing candles, work gloves and scissors from China. Bitterly disappointed, he was exploring monkhood when he met Bill Sheehan.

Sheehan had founded GRRN the previous year with the goal of fighting for a 'zero-waste' society. He was an insect ecologist by training who had morphed into an environmental activist after moving to the tiny town of Ty Ty in south Georgia, where residents were concerned the town could turn into a dumpsite for waste from other parts of the state as landfill space ran low.

King was attending a California waste-reduction conference when he chanced upon Sheehan outlining to other GRRN members how he planned to apply for a big grant from the Turner Foundation. 'These guys are nerdy,' thought King. 'They're scientists, recyclers. They don't know a thing about organising.' He butted in, suggesting that if Sheehan wanted to squeeze money out of Turner, he should go after a big-name company, like Coke. 'The Turner Foundation will want something actionable, not academic,' he told the group.

Sheehan had little experience taking on big corporations. In King he saw a force that could give GRRN the lift it needed. 'The guy was just incredibly brilliant, politically savvy,' he says. For the next few years, the two men worked closely together.

In the 1997 letter he wrote to Coke, King asked that it charge refundable deposits on containers so people would have an incentive to bring them back for recycling. The ten US states with bottle bills,[1] as laws mandating container deposits are commonly called, recycled more bottles and cans than all the other 40 states combined. Consumers brought them back to retailers or redemption centres – and if they didn't, homeless people and Boy Scouts did.

'Taxpayers and local governments presently pay the cost for disposal of your containers, which even by the most conservative estimates costs tens of millions of dollars annually,' King wrote.

*

Coca-Cola hated the idea of bottle deposits.

Although they were refundable and typically just 5 or 10 cents, Coke – like the rest of the soft drink industry – worried that deposits would hurt sales. Doug Ivester argued that consumers in bottle-bill states would drive across the border to other non-deposit states to buy their drinks.

By the 1990s, Coca-Cola and its bottlers were among the biggest opponents of proposed deposit laws. 'Bottle bills are taxation, that's what they are,' says Ivester, a Georgia native who despite travelling the world for Coke retains a strong Southern accent.

What Ivester didn't spell out when we spoke was that Coke was also afraid that retailers forced to deal with used containers would start prioritising their own beverages. 'Container deposit laws, or regulations that impose additional burdens on retailers, could cause a shift away from our products to retailer-proprietary brands, which could impact the demand for our products,' the company explained in an annual report.

---

[1] At the time, these bottle bill states were California, Connecticut, Delaware, Iowa, Maine, Massachusetts, Michigan, New York, Oregon and Vermont. Despite being called bottle bills, these container deposit laws included cans too.

Ironically, the concept of a deposit had originated within the soft drink and beer industries. Back in 1929, around 80 per cent of bottlers Coca-Cola surveyed said they used deposits to get containers back. The bottlers, many of whom were small family-owned outfits, relied on the deposits to avoid buying large numbers of glass bottles. The deposits were as high as 2 cents, which doesn't sound like much today but back then made up 40 per cent of a soda's price. Safe to say there weren't many littered bottles lying around.

Using bottlers allowed Coke to decentralise its operations and cheaply reach customers in far-flung parts of rural America. The company sent its concentrate and syrups to the bottlers, who then added still or sparkling water and sweeteners. For a long time, Coke had been a fierce advocate for returnable bottles.

Then, in 1935, disposable steel cans started making inroads. At first, Coca-Cola was lukewarm about using cans. Thick returnable glass bottles remained the cheapest option since they could be used over and over again. Its executives believed that throwaway cans would never be able to compete with returnable glass on price and would remain a small slice of the market. They also observed that Americans seemed to like drinking Coke in glass bottles; the company sold 14 billion of these a year.

Brewers, by contrast, took to the light, stackable, unbreakable cans enthusiastically, seemingly unimpressed by the glass industry's 'beer is better in bottles' campaign. Prohibition had seen the number of breweries nearly halve across America. When it ended, the three big beer companies – Anheuser-Busch, Schlitz and Pabst – saw disposables as a cheap way to ship large amounts of beer across increasingly long distances. Canned beer grew 125 per cent between 1937 and 1941, far outstripping the 32 per cent growth seen by the overall packaged beer market.

But soft drinks were an even bigger market than beer. Steel makers poured money into marketing the cans as quicker to cool and the modern choice over fusty old breakable glass. One big steel maker, Jones & Laughlin Steel, wrote letters to all its 100,000 employees and shareholders asking them to drink soda from cans to 'show off their advantages to others'. Americans watching popular TV shows like *Maverick* and *Hong Kong* were besieged by ads from Kaiser Steel urging them to drink soda from cans. American Can paid soft drink bottlers to plug canned drinks with their customers.

Coke eventually started using disposable cans in 1955, initially for export to military bases abroad. Within a decade, it was selling tens of thousands of cases of cans domestically. The first cans were steel. Later, Coke switched to aluminium, which was lighter and cheaper.[2]

Once complacent, glass container makers were alarmed. In response, they launched a thinner glass soft drink bottle. It was designed to be used once and thrown away. But, once again, many soft drink bottlers – even ones using disposable cans – were reluctant to switch to throwaway bottles. They had invested an enormous amount in cleaning and refilling equipment. Surveys showed that consumers still held the returnable glass bottles in high regard.

On the face of it, the bottlers' reluctance to embrace disposables spelled good news for the glass industry. But many glass makers believed that the days of returnable glass were numbered. Convenience, increasingly, was king and nothing was more convenient than a throwaway container. The writing was on the wall.

Convincing its soft drink customers to switch to disposable glass before can makers made further inroads became a matter of life and death for the glass industry. Its trade body, the Glass Container Manufacturers Institute, turned to Ernest Dichter, a renowned Austrian psychologist who in time would be called the 'father of

---

[2] Plastic, incidentally, quietly plays a big role in the can industry. Aluminium food and drinks cans contain an internal plastic liner, usually one containing the controversial chemical bisphenol A (BPA), to protect against corrosion.

motivational research'. Dichter had a balding head of rust-coloured hair and blue eyes that were overshadowed by thick horn-rimmed glasses. He often smoked a pipe, particularly when being photographed or interviewed – perhaps he knew it lent him an erudite air.

He prided himself on uncovering consumers' unconscious desires. When housewives wouldn't buy Betty Crocker's cake mix – despite having told market researchers how appealing the concept was – Dichter identified guilt as holding them back. He got the company to fix this by including instructions to whisk in an egg. This, Dichter concluded, was enough to make the housewives feel that they were exerting an appropriate amount of effort.

Dichter also famously encouraged Chrysler to show off the company's convertibles to draw more men into its showrooms. Although few would buy the convertibles, rubbing up against them offered a sense of freedom and excitement similar to that a mistress brought, he said. Lured by the convertibles, the men invariably chose the more sedate sedans their wives approved of.

For the embattled glass industry, in 1961 Dichter surveyed 800 American bottlers, retailers and consumers, concluding that more than 65 per cent of consumers wanted soft drinks in throwaway bottles. Americans were now buying soft drinks as regularly as salt, sugar and flour, making the returnables 'outmoded and an extraneous shopping step', wrote Dichter in a report for the glass industry.

Disposable bottles, he said, were seen as contributors to a clean and orderly household and returnables as detractors from one. 'We find that women who seem unable to achieve a fairly high degree of organisation of their household procedures leading to orderliness and cleanliness are, for the most part, guilt-ridden, defensive, and irritable within the household,' he concluded.

Apart from making women less irritable, the disposable glass bottle was also preferable to cans among both sexes. People thought it better preserved flavours. They liked that it was transparent, still 'traditional' and pleasing to the eye. The new packaging promised to increase sales – the availability of a disposable container 'heightens consumption tendencies', wrote Dichter.

Although people wanted convenience, Dichter's research also showed they harboured guilt about throwing things away. The more someone valued a product, the more likely they were to struggle with putting it in the trash. The industry would need to find ways to *convince* people to throw glass bottles away.

Dichter suggested developing a stubby bottle with a wider shoulder and lower neck than the returnable glass one in which consumers were shown to 'place high psychological value'. The glass industry's resulting disposable bottle was a huge success – so much so that once people got a taste for its convenience, they threw away returnable bottles too. Deposits had stopped keeping pace with inflation, eroding the incentive consumers once had to bring back their empties. The steady drop in return rates eventually made returnables more expensive for bottlers, setting the stage for their demise.

The final nail in the coffin for returnable glass bottles came from supermarkets. They had long fumed at the inconvenience and cost of having to accept and store returned bottles, and began pushing companies to switch to disposables.

By 1966, disposables made up 40 per cent of the soft drink market in big cities like New York and Los Angeles, and Coca-Cola was aggressively promoting their use. The shift to disposables helped the company eliminate many of its small bottlers. With fewer middlemen, Coke could now pocket a bigger slice of the profits. The number of its bottlers in the US plunged to 500 in 1979, from 1,250 bottlers 50 years earlier. By the time Lance King wrote his letter in 1997, Coke had only around 100 bottlers left in the US.

Importantly, disposability also allowed Coca-Cola and other soft drink makers to outsource the cost of dealing with their used bottles and cans onto taxpayers, since these immediately became municipal waste. The drinks industry put the extra dollars to good use, pouring money into ads aimed at getting people to drink more. Between 1955 and 1969, the per-capita consumption of soft drinks in the US soared by 88 per cent.

Inevitably, litter from soft drink containers piled up. By the early 1970s, disposable drinks containers comprised nearly one in five items found littered on highways. Predictably, a backlash ensued. More than 350 bills to ban, tax or slap a deposit on drinks containers were introduced in Congress, state legislatures and local jurisdictions.

Keep America Beautiful – a non-profit whose founders included Owens-Illinois, a major Coca-Cola packaging supplier – took to emphasising the responsibility consumers had to not litter. It paid for a, now-infamous, TV ad in which an actor dressed as a Native American watches a passing driver fling a bag of trash out of the window of his car. A single tear rolls down the actor's face. 'People start pollution,' says the voiceover. 'People can stop it.'

Against this backdrop, Coca-Cola – even as it pushed deeper into disposables – put out an ad just before the first Earth Day in 1970, touting the returnable glass bottle as its favoured container. 'This is the bottle for the Age of Ecology,' proclaimed the full-page ad. 'The reusable, returnable bottle for Coke is the answer to an ecologist's prayer,' it added, somewhat hyperbolically.

The ad went on to explain that Coke's glass bottles made around 50 round-trips, which the company explained meant 50 fewer chances to add to the world's litter problem. 'Like the ecologists, we prefer it over the other containers,' it said. 'So buy Coca-Cola in returnable bottles. It's best for the environment – and your best value.'

*

The Age of Ecology advertisements ran for several years. They were doubly ironic given that, behind the scenes, Coca-Cola was readying the disposable drinks container that would pose the biggest waste challenge of them all: the single-use plastic bottle.

In 1969, the year before the ad first ran, Coke had begun quietly experimenting with plastic, teaming up with Monsanto to create a 10-ounce bottle made from the chemical giant's Lopac plastic.

Plastic bottles had first burst onto the scene in late 1946. They quickly caught on among the makers of cosmetics, liquid deodorants and medicines. The industry got another shot in the arm when P&G, Colgate-Palmolive and Unilever began putting their laundry detergents in plastic. By 1960, plastic bottles were cheap enough to compete toe-to-toe with glass and tin, and were being used by the makers of everything from lemon juice to glue. But it took several more years until engineers found a way to create a plastic bottle that could withstand the intense carbonation of fizzy drinks.

By March 1970 – when the Age of Ecology ad was being debuted – both Coke and Pepsi were test-marketing plastic bottles for soda; Pepsi's bottles were made from Barex resin produced by Standard Oil of Ohio (Sohio). Monsanto and Sohio argued that the plastic containers would *help* alleviate America's growing solid-waste problems. They were lighter than glass and, according to the companies, emitted the same amount of pollutants as paper when burned. Sohio sent 5,000 letters to government officials describing the plastic bottles as a 'new product that can solve the litter problem'. The company's vice president said the bottles burned to a fine ash. He also claimed that, when buried, they appeared to return to their basic molecules of carbon, nitrogen, hydrogen and oxygen.

Sohio even told journalists it was testing a plastic bottle that would 'self-destruct' within a year of being emptied, and that its chemicals would be absorbed by the soil. The wild claims attracted positive press. 'A new plastic bottle designed to rid America's landscape of unsightly soft drink and beer containers has been developed,' wrote a reporter for a local Ohio-based newspaper.

Both Coke and Pepsi saw plastic bottles as the perfect way to sell ever larger quantities of soda. Large glass bottles were heavy and, when refrigerated, condensation made them slippery and prone to breaking. Market research showed that consumers would choose plastic bottles over glass as long as they weren't more expensive.

In 1975, Coca-Cola launched the Easy-Goer, a 32-ounce green-tinted plastic bottle made from Monsanto's acrylonitrile polymer. The bottle was 17 ounces lighter than a similarly sized glass bottle. The name said it all: the bottle was designed to make drinking soda easy, particularly for consumers on the go.

As Coke worked to ensure Americans bought bigger and bigger bottles of soda, the company also moved to normalise their new habits. It stripped the words 'king', 'giant' and 'family' from its package descriptions, instead simply listing how many ounces the containers housed. In the 1980s, Coke launched a 3-litre plastic bottle, the Thirst Buster. Not content with that, it also began test-marketing a 4-litre plastic bottle named the Super Thirst Buster.

But, from the beginning, there were concerns about the drinks industry's use of plastic. It wasn't just the inevitable litter, or that the bottles were made using finite oil and natural gas resources. Environmental groups said the bottles could cause serious health problems.

In April 1976, the National Resources Defense Council sued the Food and Drug Administration to halt the use of plastic soft drink and beer bottles. The non-profit raised concerns about Coke's use of acrylonitrile, a close relative of vinyl chloride which was shown to cause cancer in animals. Studies had shown that rats fed acrylonitrile for three weeks developed ulcers and damage to their adrenal glands. Japanese workers exposed to it showed abnormalities in their liver functions. NRDC maintained that acrylonitrile and hydrogen cyanide could migrate from the plastic into drinks.

At the time, Coke dealt with the concerns by sticking with plastic but switching to a different type. PET, developed by Nathaniel Wyeth, a DuPont engineer, became the industry standard and remains so today.

*

By the early 1990s, most of Coke's drinks across the US were being sold in disposable cans and PET bottles. Returnable glass made up less than 1 per cent of what it sold in the US. Yet the company still wanted to hang on to the wholesome image evoked by its 'bottle for the age of ecology'.

Coca-Cola's advertising sought to evoke nostalgia, and for this the company used imagery of its old hobble-skirt glass bottle, often ice-cold with beads of condensation dripping down it. It had used the distinctive bottle since 1916. Also called the 'Mae West' bottle after the American actress's famously curvaceous figure, the shape was arguably the most recognised corporate symbol in the world. Doug Ivester was convinced that Coke needed to keep the contour bottle high in the public's imagination.

Hailing from New Holland – a small village in Georgia built for cotton mill workers – Ivester had started out life with modest ambitions. As a young teenager, he had worked the cash register and carried bags for customers at the Kroger store in Gainesville. He had longed to buy a swanky car like the 1964 Pontiac GTO driven by a regular customer. The man told Ivester that he was a certified public accountant. That car spurred the future Coke CEO to get an accounting degree.

'The bridge from growing up in that modest environment to having what I considered to be a good income-producing job was a CPA,' Ivester says. 'I knew some people who had done it. They were successful and happy, they had good lives and didn't need to be multimillionaires.' Ivester spent a decade at an accounting firm, Ernst and Ernst, where much of his time was taken up by Coke, his main client.

In 1979, he joined the drinks giant's auditing department, putting in the same long hours he had at Ernst. His hard work paid off; six years later, at the age of 37, he became Coke's chief financial officer. Then, in 1989, the company's Cuba-born CEO Roberto Goizueta appointed Ivester as Coke's head of Europe. His time abroad was short-lived. While Ivester was still apartment-hunting during his first year in Europe, Coke's US president Ira Herbert started having heart problems. One day, Goizueta phoned Ivester in his hotel room to say Herbert wanted to retire. 'I need you back here to take Ike's place,' Goizueta told him.

Soon after returning to Atlanta, Ivester made it his mission to renew Coke's focus on the contour bottle, convinced it would help sales. While the company was doing brisk business overseas, growth at home was sluggish. 'I was concerned about the loss of brand imagery,' remembers Ivester. 'I felt the contour bottle was a major conveyor of image and quality. It conveyed heritage and all the right things.'

He charged a 29-year-old marketing executive, Susan McWhorter, with figuring out whether Coke could make a contour plastic bottle. 'You can segregate people into two groups. There's the problems group – that says let me tell you all the problems we're facing – and the solutions group,' says Ivester. 'Susan was a solutions person.'

McWhorter had gone to Ivester's alma mater, the University of Georgia, and had long idealised her now employer. As a child, she'd spend all week looking forward to the icy bottle of Coke her grandmother gave her every Saturday at 11am. At age seven, McWhorter's payment for sweeping up hair at Wilma's beauty parlour in her hometown of Ocilla was a 6.5-ounce returnable bottle of Coke.

McWhorter polled consumers who indicated they'd prefer contour bottles over straight-walled ones by a margin of five to one.

Younger consumers saw the bottle as modern and different, while older people who remembered the shape associated it with quality.

Consumer interest confirmed, Ivester wanted not just a plastic replica of the glass bottle, but a much larger version. Over the years, Coke had steadily increased the sizes of its fountain drinks. A large soda now stood at 20 ounces, a full 4 ounces bigger than the previous iteration. 'We were really training consumers at that time to drink more and more,' says McWhorter.

Coke didn't have to charge consumers much more because the profit margin on fountain soda was so much higher than on bottles and cans. The huge fountain sodas paved the way for the company to slowly but surely reshape consumer expectations, creating a thirst among Americans for larger amounts of soda, across every packaging type. 'The consumers just ordered a large,' explains McWhorter. 'They didn't know whether it had 16 ounces or 20 ounces in it. For us, the thinking was, "We sell more, we make more – so let's size up."'

Ivester instructed McWhorter to find a way to make a 20-ounce plastic bottle that looked like the original 6.5-ounce glass one without compromising the design's integrity. But sizing up was looking expensive. For one thing, the bottle manufacturers needed to use extra plastic to give the curvy bottle added reinforcement. Curvy bottles also couldn't be blown as quickly as straight ones. The bottles wobbled on filling lines. Bottlers trialling them pumped in just 10 per cent of the liquid they usually filled in a day. Modifying the filling equipment could cost a bottler between $1 million and $2 million. Yet the payoffs were uncertain and Coke's recent track record was anything but reassuring.

Ivester had only recently pushed bottlers to get behind a clear, sugar-free soda intended to compete with a clear, colourless Pepsi. Coke called it Tab Clear and Ivester told reporters the product would be 'marketed for what it is – a study in contradictions'. But the marketing confused people. They didn't recall the ads. The key message – that Tab Clear had a 'mysterious flavour' – didn't resonate. Tab Clear's many critics said it looked like lemonade but tasted like weak cola. By late 1993, a year after Tab Clear's launch, Goizueta was warning that the product was dead.

'We were going into an environment where bottlers said, "Yeah, we've seen all this research from the company on how great things are going to be. We've done all this work but it doesn't pay off",' says McWhorter.

To win over the bottlers, Ivester knew he needed to put his money where his mouth was. 'Coke will loan you the money for the conversion of your lines,' he told them. 'If you execute the marketing plan Susan and her team give you and don't meet the target numbers, I'll forgive those loans.'

Just as P&G's chairman had, in 1962, taken a punt by ploughing money into making big quantities of Pampers so he could slash prices to 6 cents apiece, Ivester too was taking a big gamble. The company was ploughing tens of millions of dollars into modifying the bottlers' lines, which meant the amount of soda it sold needed to jump significantly to cover the added costs.

In January 1993, Coke launched the plastic contour bottle in test markets in Alabama and Tennessee. Sales jumped 25 per cent. 'For more than 75 years, our contour bottle design has been the unparalleled symbol of quality,' said Ivester, announcing the launch. 'The new 20-ounce package preserves that heritage while offering the convenience of a recyclable plastic package to today's consumer.' The plastic contour bottle, the company told investors, will 'invite consumers to drink more Coca-Cola, more often and in larger sizes'.

The *Wall Street Journal* dismissed the bottle as a 'marketing gimmick', writing 'New 20-Ounce, Plastic Version Is Trying to Be Nostalgic and Hip at the Same Time'. Unsurprisingly, Pepsi denigrated the bottle, telling the *Journal*: 'The more nostalgic Coke gets, the more Pepsi looks progressive, from an image standpoint.'

But Coke saw the potential for its new drinks packaging to be far more than just a container. 'We call it unleashing a powerful marketing tool that touches consumers where our competitors cannot – in the palms of their hands,' the company's chief marketing officer Sergio Zyman told a trade magazine.

Retailers loved the new bottle, handing over large chunks of shelf space to Coke. Profit margins on single bottles of Coke were significantly higher than on ones that came in packs. The contour bottle was aimed at growing this single-serve segment, targeted at people on the go. Consumers loved it too. Coke's sales volumes jumped as much as 90 per cent in parts of the US where the bottle had been launched. The results were beyond anything the company had expected.

By September 1994, Coca-Cola told investors that it was on track to post its largest sales volume of any quarter in the past five years. Ivester rolled the bottle out nationally. Coca-Cola forecast that total sales of 20-ounce Coke in plastic bottles for that year would jump 50 per cent from the year before when only straight-sided bottles were offered. 'Contour is a brand in and of itself,' boasted Ivester.

Through the 1990s, the returnable glass bottle's already tiny share of all carbonated soft drink sales in the US dwindled. By the decade's end, it stood at 0.2 per cent.

*

By the late 1990s, Coca-Cola was selling an estimated 20 million sodas each day in plastic bottles in the US. That translated into nearly 14,000 plastic bottles a minute. The recycling rate for plastic bottles in the U.S. was under 24 per cent.

Environmentalists were infuriated. The biggest problem, they said, wasn't that consumers weren't recycling, rather it was that companies weren't buying old plastic to make new packaging. Putting something in a recycling bin is meaningless if nobody wants to make a new product out of it. In 1998, the theme for America Recycles Day was 'if you're not buying recycled, you're not really recycling'.

Lance King of GRRN saw pushing for Coke to use recycled plastic as a gateway to holding companies financially responsible for dealing with the waste generated by their products. At its heart, this was the concept of 'extended producer responsibility' – the same one Carl Lehrburger had proposed when he questioned who should bear the cost of discarding disposable diapers.

Targeting Coke had helped GRRN attract $130,000 in grants from foundations associated with media mogul Ted Turner and journalist Bill Moyers. The non-profit's campaign included five different ads. They ran in the *New York Times* and later also in the *Wall Street Journal*.

One carried a note of Ivester's 1998 compensation, alongside the calculation that keeping his promise to use recycled plastic would cost no more than the stock options Coke had granted its CEO that year, or 0.5 per cent of the company's profits.

GRRN encouraged consumers to mail empty plastic soda bottles to Doug Ivester with the message 'Take it back and use it again'. In 1998, activists dropped off a giant 3.5 by 7 feet 'buy recycled' pledge form at Coke's Atlanta headquarters.

Ivester paid them little attention. 'The fleas come with the dog,' he says stolidly when I ask whether the GRRN ads bothered him. 'You have those kinds of jobs, they're great jobs but this kind of thing comes with the territory.'

By now, Ivester also had bigger reputational problems than the ones posed by GRRN's attacks. Goizueta had died of lung cancer and Ivester had been hastily parachuted into the role of Coke's CEO in October 1997 following a board meeting that took all of 15 minutes. To say Ivester's time at the helm since then had been bumpy is an understatement.

Think before you drink Coca-Cola.

Coca-Cola broke its promise to use recycled plastic in the *10 billion* bottles it pumps out each year.

Coke® has 45 percent of the domestic soft drink market. So until Coke starts buying recycled plastic for its bottles, plastic recycling is stalled.

Coke CEO Ivester already knows recycled-plastic bottles are better for the environment. He said so when he pledged in 1990 to use them.

Coke's claim to be "environmentally responsible" is called into severe question by his broken promise on plastics. And for the past 30 years, Coke has lobbied behind the scenes *against* effective recycling legislation.

All this should make us *think before we drink* another Coca-Cola product...

• *Think* about the billions of plastic Coke bottles clogging landfills and millions more littering our beaches, parks, roads, and farms.

• *Think* about the waste and pollution from manufacturing virgin plastic Coke bottles instead of using recycled plastic.

• *Think* about the 30 percent drop in plastic soda bottle recycling rates since Coke broke its promise.

Bottom line? Coke should own up to the waste it creates.

The answer is easy. Coke made a promise. They should keep it. It's a message even a child can understand.

Call Coke at (800) 571-2653. Tell them to honor their pledge to use recycled plastic.

Coke CEO M. Douglas Ivester

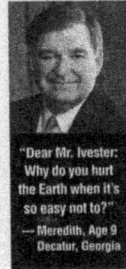

"Dear Mr. Ivester: Why do you hurt the Earth when it's so easy not to?" — Meredith, Age 9 Decatur, Georgia

Coke's waste and pollution— a legacy for our children?

GRRN GrassRoots Recycling Network

Learn more from the GrassRoots Recycling Network, P.O. Box 49283 Athens, GA 30604. **www.grrn.org**

PAID FOR BY THE FLORENCE FUND

Coke, Coca-Cola & Coca-Cola Classic are registered trademarks of The Coca-Cola Company.

Image courtesy of Bill Sheehan

Schoolchildren in Belgium had fallen sick after drinking Coca-Cola products, sparking the largest recall in company history. Coke was facing a lawsuit from African-American employees who said they were discriminated against on pay and performance evaluations.

Ivester had also badly bungled an attempt to buy Orangina – a sparkling orange drink owned by Pernod Ricard – in the face of staunch opposition from France's competition authorities.

With their CEO in the firing line, Coke's executives – whose strategy so far had been to ignore allegations that it was failing to use recycled plastic – broke their silence. 'We conducted a test, and the results were not economically sustainable at that time,' became the party line, spouted by Coke's spokespeople. 'We did not break a pledge.'

The company told concerned consumers that soft drink containers were America's most recycled package, with 58.7 per cent of these being recycled. Of course, it wasn't aluminium cans and glass bottles that consumers were concerned about.

Coke's customer service began telling people phoning in that the company used recycled plastic on a regular basis. But non-profits estimated that recycled plastic amounted to just 0.1 per cent of the PET that Coca-Cola used to make new bottles each year in the US.

The National Soft Drink Association denigrated environmentalists for asking Coke to use more recycled plastic. 'Our position is that the use of recycled plastic is costly and that environmentally questionable mandates are not needed,' the influential trade body told reporters. 'Right now it is not cost-competitive to use recycled plastics.'

GRRN kept up the pressure. In January 2000, three days before the Super Bowl was due to kick off in Atlanta, King and his colleagues unleashed a 20-foot-high inflatable replica of a plastic Coke bottle at a press conference. The message attached read: 'Demand Coke Use Recycled Plastic Always'. Coke was a major Super Bowl sponsor and the gathering – held at Woodruff Park in downtown Atlanta, named after a legendary former Coke president – was designed to ensure visitors to the city would learn about the company's broken promises.

The bottle was regularly couriered from one city to another for use in various protests. On one such journey after the Woodruff protest, it failed to arrive at its destination. It was never seen again.

'We were always conspiracy-minded about what happened,' says Bill Sheehan from GRRN. 'Maybe the other side got to it.'

*

In May 2001, Coke's environmental manager Ben Jordan walked into a conference room at the Hilton hotel near Atlanta airport. He took his seat at a large table alongside Lance King and other environmentalists, recyclers and state environmental officials.

A handful of other companies were represented too. There was Waste Management – America's largest waste company – and Tomra, whose reverse vending machines were widely used in Europe to take back used bottles for deposit programmes. Also at the meeting were top executives from Coke's plastic bottle maker, Southeastern Container, and from Beaulieu, a major carpet maker that wanted more used plastic bottles to make synthetic carpets.

The discussions Jordan had joined were the start of a collaboration that had the stated goal of doubling the US recycling rate for drinks containers made from plastic, aluminium and glass to 80 per cent. It sounded hugely ambitious, but the rate wasn't far from the average that states with container deposit programmes were hitting.

King had named the group Businesses and Environmentalists Allied for Recycling, or BEAR for short. Its eclectic nature made it notable. 'It was the first time we had so many different stakeholders at the same table,' recalls Steve Navedo, a PET recycler who was one of the participants. Jordan's presence gave the discussions added legitimacy, adds Navedo.

The collaboration was led by a former Coke executive called Pierre Ferrari. A native French speaker who was born in the Belgian Congo, Ferrari studied economics at the University of Cambridge and first joined Coke in Johannesburg. He worked at the company for two decades until, one Saturday morning in 1995, he had wandered into the kitchen of his Atlanta home, still bleary-eyed, only to find his 13-year-old son washing down a cold slice of pizza with a can of Coke. 'What the hell are you doing?' demanded Ferrari. 'Why don't you eat a healthy breakfast?'

'Dad, all you do every day is sell Coke to everyone, so why can't I drink one?' retorted his son. The exchange became a turning point in Ferrari's life. It sparked weeks of soul-searching and eventually led to his decision to resign. 'I was disenchanted with the product, its sugar, the impact on society and obesity,' he says bluntly.

Many of BEAR's members thought that levying deposits on drinks containers was the best way to raise recycling rates. Coke, unsurprisingly, had refused to join the group. But with Ferrari acting as a mediator, it agreed to participate in BEAR's recycling research.

A few weeks earlier, Coke's shareholder meeting had heard a proposal from a group of shareholders holding $50 million in stock. They wanted the company to use more recycled plastic and increase recycling rates. The proposal failed to garner enough votes but, internally, some senior Coca-Cola executives were starting to feel that the company needed to do more to appease its critics.

At BEAR's first meeting Ben Jordan showed up informally dressed. He had a southern drawl, a relaxed manner and didn't know a whole lot about plastics recycling. He had worked for Coke for fewer than five years and the other men in the room quickly warmed to him. 'I was kind of surprised that Coke let Ben be part of it,' says Navedo, the PET recycler. 'But there he was.'

The men at the Hilton meeting agreed they'd assess the costs and effectiveness of various plastic bottle collection options in the

US to see which would most effectively raise recycling rates. The discussions didn't always go smoothly. 'There was definitely tension at every meeting,' recalls Ed Boisson, the project's lead researcher. 'The industry at all costs wanted to avoid deposit systems and although Ben Jordan came in with an open mind, it was obvious he was under pressure.'

After eight months of meetings, in January 2002 the BEAR group put out a 112-page report warning that drinks container recycling rates were set to 'steadily decline' if things continued as they were. The report – pulled together by a collection of consulting firms seen as representing both industry and environmentalists' viewpoints – concluded that *financial incentives* were the best way to increase recycling rates.

The firms' number-crunching showed that, across the ten states that charged deposits, the rate at which containers were recovered for recycling was 78 per cent. In states with no deposits, the rate was just 18.5 per cent. While collecting and cleaning old containers meant deposit programmes looked the most expensive option, the report explained that counting the deposits that consumers didn't claim as revenue would hugely defray the cost, making container deposit programmes less expensive than municipal kerbside ones.[3]

While the report didn't overtly endorse using a deposit programme to increase recycling rates, the implication was clear: charging refundable deposits on drinks containers was the way to go. Jordan signed the report, his signature appearing prominently towards the top of the first page. The Container Recycling

---

[3] Kerbside recycling refers to local government-funded collections from individual homes and apartment buildings. It's often 'single-stream', which means people can put all their recycling in a single bin or bag rather than needing to separate it into glass, metal, paper and plastic.

Institute, a non-profit that had participated in the meetings, rushed out a press release. 'The report provides hard numbers that confirm the superior performance of deposit/return programs over curbside recycling and drop-off collection,' it crowed.

Nothing could have prepared Ferrari and the other group members for the backlash that followed. 'We could not disagree more strongly with the BEAR report,' said William L. Ball III, the president of the National Soft Drink Association. 'The numbers on which the study is based are, in many cases, so flawed that they make it impossible to draw any meaningful conclusions from the data.' NSDA's environmental affairs head went even further than his boss, suggesting that 'the numbers may have been manipulated to achieve a desired conclusion'.

Northbridge Environmental, the same consulting firm that had defended plastic foam makers, quickly came up with a counter analysis. It alleged that BEAR had underestimated the costs of collecting used containers via deposit programmes and overestimated the costs of kerbside recycling.

Coke did an abrupt U-turn. It yanked Jordan off the project, which had been slated to move on to a second phase intended to culminate in an agreement on what the best recycling system was and how to implement it.

After Jordan walked away, relations between Coke and the environmentalists soured. 'Whether participation in BEAR was an honest effort on Coke's part or just another delay tactic, I can't say,' says Sheehan from GRRN. 'Clearly, scuttling the effort bought them 20 years of profits from continued off-loading responsibility onto taxpayers and, unfortunately, two decades of continued staggering plastic bottle waste.'[4]

---

[4] Jordan and Coca-Cola declined to offer an on-the-record comment about BEAR for this book.

By the end of 2002, the PET bottle recycling rate in the US had dropped to less than 20 per cent. The rate was half that achieved in 1995 and marked the seventh consecutive year of decline. Despite Coke's proclamations to its shareholders, plastics recycling was going backwards.

With Coke out the door, the BEAR effort to raise recycling rates collapsed. 'It just evaporated because without the soft drink companies' support, you can't really get anything done,' says Ferrari. 'If we can't convince them with an economic argument, we can't convince them at all.'

CHAPTER 11

TAKEN CAPTIVE

'Some would say that Coke infiltrated NRC'

When Kate Krebs stretched across her enormous desk at the National Recycling Coalition's headquarters to answer the phone one morning in 2001, the unfamiliar voice on the other end identified itself as belonging to Scott Vitters. Vitters introduced himself as Coca-Cola's environmental affairs head. Upon hearing this, Krebs sat a little straighter in her chair. 'What can I do for you?' she asked.

Krebs was a long-time recycler who had recently moved from the small progressive town of Arcata in northern California to Alexandria – a city on the banks of the Potomac River, just south of Washington, DC – to become deputy director for NRC.

NRC was the same body that had unsuccessfully attempted to negotiate with the Society of the Plastics Industry to overhaul the chasing arrows symbol back in the early 1990s. It had been founded in 1978 and, by the early 2000s, had thousands of members, among them state recycling organisations, municipal recycling coordinators, private recyclers and waste-reduction non-profits.

Coke had had little contact with the recycling organisation until now. Vitters was helming one of the most high-profile recycling efforts the company had ever been involved in – and he needed help. Having asked around, it seemed Krebs could be the one to provide it.

The Atlanta drinks giant was a sponsor of the upcoming 2002 Winter Olympics in Salt Lake City, Utah, and had been a major backer of the Olympics for 74 years. Backing events that emphasised physical activity had become a common marketing tactic for an industry whose message to an increasingly obese America was that inactivity, not sugary soda and snacks, was the primary cause of their growing waistlines. Vitters was out to ensure that the publicity Coke received was nothing but glowingly positive.

But, already, the upcoming event had been mired in scandal, triggering multiple crisis meetings inside Coke. It emerged that organisers lobbying for the Olympics to be held in Utah had given $780,000 in payments and gifts to members of the International Olympic Committee. News outlets reported that the payments were used for plastic surgery, scholarships and even sex. Many sponsors had fled. Coke wavered but ultimately held steadfast. The soft drink maker's research showed that being strongly associated with the Olympics had helped its brands gain people's respect, and by extension their loyalty. Coke's products were sold in more than 200 countries and, apart from the FIFA World Cup, the company's executives saw the Olympics as the only truly global sporting event – an unmissable opportunity to market its brands.

Having barely survived the payments scandal, the Olympics' organising committee was now being targeted by non-profits over the large volumes of waste the event was expected to generate. The committee – newly led by Mitt Romney, who'd soon go on to become the governor of Massachusetts and who had been brought in to clean things up – was on a tight budget. Given this, it planned on having people dump all trash and recyclables into just one bin.

This had riled the GrassRoots Recycling Network, which argued that mixing used bottles and cans with food waste, leftover chewing gum, soggy napkins and coffee cups spelled disaster. While the Salt Lake Olympics Committee was touting the event as the 'Greenest Games Ever', GRRN drily pointed out that this would not be the case if recycling rates were rock bottom.

Vitters was all for the idea that recyclables needed their own receptacles. A trim man who spoke quickly and confidently, Vitters had joined Coke in 1997 after working as a consultant on waste issues for a US Environmental Protection Agency contractor. He saw the Olympics as a golden opportunity for Coke to execute a recycling programme that showcased its leadership on waste issues. At his urging, Coke ponied up the money for 10,000 new trash and recycling bins in and around the sporting venues. The company spent $300,000 on developing recycling and composting programmes in Salt Lake City during the Olympics and committed to buying 100,000 pounds of plastic recycled from drinks sold at the games.

But there was another event that Vitters needed to manage the waste disposal for – the relay that would transfer the Olympic flame via icicle-shaped torches from Atlanta to Utah. The torch relay would begin two months ahead of the Olympics in early December 2001. Coke had kicked in $7 million to sponsor the event alongside Chevrolet and for months had been marketing the hell out of it.

The company had sought nominations for thousands of regular Americans to be torch-bearers, hiring Lance Armstrong to be the face of an advertising blitz that reached 85 per cent of the country. The cyclist and three-time Olympian appeared on movie theatre screens and vending machine covers across the US to promote the relay. Armstrong was hugely popular at the time – this was long before he'd be implicated in a doping scandal – and his involvement drew 210,000 torchbearer nominations, many times more than the previous games. From these, 11,500 people were chosen to run relay legs of 0.2 miles each.

The 3-foot-long glass and silver torch would be lit by boxer Muhammad Ali from a huge cauldron at Atlanta's Centennial Olympic Park, signalling the start of the flame's journey across the country. After the flame left Atlanta, it would travel across the US, stopping in two cities a day while covering 13,500 miles through 46 states. Finally, it would arrive in Salt Lake City in time for the start of the Winter Olympics on 8th February, 2002.

Every afternoon during its 65-day journey, there'd be a street party in a city or town. Every night, there'd be an even larger one in another location, courtesy of the torch relay's $60,000 party budget. In each location, party-goers would get free drinks courtesy of Coca-Cola.

After months of looking for ways to get bottles collected in all 130 cities where Coca-Cola would be handing out drinks, Vitters had realised this would prove very complicated and hugely expensive. And so he phoned Krebs to ask if NRC could help. He explained how Coke would need recycling bins but also coordinators to form 'green teams', shepherding people towards the bins and keeping waste in check. He knew that NRC had a large network of recyclers across every state and that those bodies in turn knew recyclers in every town. 'Can you help us?' he asked Krebs.

'You're going to have to pay for it,' she told him. 'But it sounds cool.'

*

By tying up with Krebs, Vitters had found Coke a valuable new ally in the recycling community. Blonde, blue-eyed and respected by recyclers up and down the country, Krebs was raised by conservative Catholic parents in Ventura County, California. As a young girl, she spent hours by the water, often going out on her father's small sailboat.

Then, in 1969, an offshore oil well exploded, sweeping 250,000 gallons of oil along 30 miles of coastline near where she lived. It was the largest oil spill ever to have occurred in US waters at the time. Stumbling over the bodies of suffocated birds and other animals left a lasting impression. Krebs, then a teenager, took time off high school to help with clean-up efforts. 'I had lived a pretty wonderful life until that point,' she says. 'To see seabirds and fish dead on the beach was shocking.'

An environmental seed was planted. When Krebs finished school, she enrolled at Arcata-based Humboldt State University, known for its environmental programmes. At the time, Arcata had just started a drop-off recycling centre – one of the nation's earliest – and Krebs began managing the centre's payroll and accounting.

The Arcata operation was very much a labour of love. The centre was 400 miles from the nearest end market for recyclables and any money earned was erased by fuel costs. Started by conscientious objectors to the Vietnam War who were made to do community service, the centre was housed in a parking lot and was staffed largely by local volunteers, including students and former convicts who needed job skills and training. 'It was very community based, very grassroots, do your part to help save the world,' remembers Krebs.

Krebs – like everyone else at the recycling centre – was a strong advocate for a state container deposit law, the kind Coke despised. She would sometimes drive to the state legislature to sit in on hearings for California's proposed bottle bill. When, in 1986, California eventually implemented a container deposit law, people came to the Arcata recycling centre to drop off bottles and cans and get their money back. Krebs saw this as an opportunity to get them to spend their recovered nickels and dimes; she opened a building material reuse yard and a clothing thrift store.

Home to groves of ancient redwoods, Arcata was an odd place. It had long been a mill town, home to a training school for the timber industry, but since the 1960s its culture had been increasingly shaped by students and San Francisco migrants looking for a more laid-back lifestyle. Many of these newer arrivals doubled as anti-logging activists and sometimes they clashed with the loggers. When Krebs ran to be Humboldt County Supervisor, her opponent was 'Chainsaw Annie', a local woman nicknamed for her ferocious advocacy for the timber industry.

'Families of loggers would parade through Arcata and environmentalists would do the same,' recalls Ed Boisson, who lived in Arcata in the 1980s as a graduate student and became friends with Krebs. 'It was jobs versus trees.' Activists used to put chainsaw-destroying spikes in redwoods to protect them, sometimes harming mill workers. The activists were occasionally peppersprayed by supporters of the loggers.

Although Krebs lost the county supervisor race, working in Arcata helped her develop a knack for diplomacy and playing both sides of the room. Recycling, she recalls, was the one goal everyone – regardless of political leaning, age or career – seemed to agree was worthy. She eventually became executive director of the Arcata centre and also a member of a non-profit called Californians Against Waste.

Everyone in California's environmental circles knew Krebs. 'She was a pillar of the Arcata business community running the recycling centre,' says Boisson, who went on to become an environmental consultant and was hired to work on the BEAR recycling report.

In the 1990s, Krebs became involved with NRC, at first serving as the head of its rural recycling committee. She got on well with its then director, Will Ferretti, who in 1999 asked her to move to Alexandria to be his deputy. 'I needed the extra bandwidth and Kate had high credibility within the recycling community,' remembers Ferretti.

But the tie-up was short-lived. In September 2001, only a few months after Vitters had phoned her to ask for help, Ferretti resigned and Krebs was named executive director of NRC.[1] The role of executive director made Krebs the sole official spokesperson for the most notable and diverse recycling advocacy organisation in America. She was also suddenly the person in charge of managing its budget.

As a teenager, Krebs had helped out with the accounts for a family real estate and cattle brokerage business, and once at Arcata had been in charge of the recycling centre's finances. Not everyone thought this was a good fit. 'I think it's fair to say finance and accounting were certainly not her strong suit,' says Mark Loughmiller, who took over from Krebs as the Arcata centre's executive director and recalls how large cheques were held for months and expenses outstripped revenues. 'I chalked it up to sloppiness.'

When Krebs became executive director of NRC, the organisation was already short on money. George W. Bush had been named president and, according to Krebs, funding from the Environmental Protection Agency had all but dried up. Then the September 11 terror attacks happened.

Krebs, already a nervous flyer, had been on a plane heading west from Washington, DC when the hijacked planes crashed. The pilot on her United flight offered no explanation when he announced they'd be making an emergency landing. A few minutes later, Krebs was bending over and holding her knees as the plane skidded down a short runway somewhere in Michigan. There were no televisions at the airport and she had few details about what had happened. She only knew that it was significant and that there would be no flights to take her back to DC for several days.

---

[1] For a few months, Krebs was interim executive director and, by early 2002, she was executive director.

Krebs joined the long line winding towards the Hertz rental desk, getting to the front in time to claim the last available vehicle – a nine-seater van still muddy from the kids' soccer team that had last ridden in it. 'Anyone know the route to DC?' she asked the people behind her, dangling the keys. 'I got space for a few more.'

Krebs piled eight other people into the van. The group hardly spoke the whole ride back. By the time Krebs eventually made it to her Alexandria apartment, it was 2am. Exhausted, she turned on the TV to watch images of smoke billowing from what had once been the World Trade Center. It was only then that the enormity of what had happened sank in.

Two weeks after the planes crashed, however, Krebs was back at the DC airport, scuttling past armed policemen so she could fly to Seattle. The September 11th attacks had forced NRC to cancel its big money-maker for the year – its annual recycling conference, originally due to be held in Seattle at the end of the month. Krebs's task was to negotiate with the hotels booked for the conference. 'The hotels kept saying it was an act of God so we had to pay,' she remembers. The convention centre was also threatening enormous cancellation fees.

'The year after 9/11 was complete chaos,' Krebs says. 'We were hanging on as an organisation by the skin of our teeth.'

In the wake of the terror attacks, Krebs felt a rising pressure to find new funding. The majority of NRC's members were local municipal recycling coordinators. They paid just $20 in fees each year but were continually pushing the organisation to do more to improve recycling. They had 'huge expectations', says Krebs. 'What we really needed were private-sector dollars. NRC hadn't made this leap yet because there was still tension between the corporates and environmentalists.'

*

Working closely with Coke gave Krebs her first real taste of Corporate America. She saw the potential to take things further.

The Olympic torch relay, which kicked off fewer than three months after the September 11th attacks, had been a great success. Coke had recruited family members of men and women who had died aboard the hijacked planes to carry the flame. Its marketing ensured that tens of thousands of Americans turned up to watch the torch pass through their cities. 'As we got closer and closer to Utah, the crowds got enormous, almost to the point of scary,' remembers Julie Seitz, a Coke executive who poured two years into ensuring the torch relay[2] ran to plan. 'In Utah, it was millions of people, 20 deep everywhere you looked and it got pretty crazy.'

Coke hired 80 recent college grads to hand out free drinks from refrigerators kept on trucks that drove a quarter mile ahead of the relay. They got the crowds swaying to music pulsating from large speakers mounted on the trucks, while the bins and recycling volunteers Krebs had helped organise kept waste and litter under control. Vitters walked away from the relay with his head held high. He made sure to keep in touch with Krebs.

In 2003, Krebs made an announcement that surprised many NRC members. She would be facilitating closed-door meetings between Coca-Cola, Anheuser-Busch, Coors Brewing, Miller Brewing, Heineken, Pepsi and Nestlé Waters. Similar to the discussions that BEAR had had the previous year, the new group – called the Beverage Packaging Environmental Council – would look for ways to raise container recycling rates. However, unlike BEAR's notably diverse group of participants, BPEC would consist entirely of representatives for the drinks companies, with the

---

[2] The torch relay is a bit of a misnomer since it was the flame being passed from torch to torch.

exception of Krebs. Other NRC board members were told that, due to the sensitive nature of the talks, they couldn't be part of them.

The companies behind BPEC, unsurprisingly, were against bottle bills from the get-go. Many of them belonged to the National Soft Drink Association (which by now was called the American Beverage Association), the body that had slammed the BEAR report's recommendation of financial incentives being the best way to raise container recycling rates. By contrast, NRC's official policy was to be neutral on container deposit programmes, given that its members hailed from many different states, several of which had bottle bills.

Krebs's involvement in BPEC raised eyebrows among some of NRC's members. They thought the recycling body shouldn't be getting so closely involved with the drinks companies considering the latter's vested interests in ensuring the costs of recycling remained with taxpayers. 'BPEC made people very concerned,' recalls Mark Lichtenstein, a one-time NRC president who led the coalition's negotiations around the chasing arrows symbol, and who remained involved with the recycling body through the 2000s. 'It was about the cosying up of NRC to the American Beverage Association.'

Krebs, ever the pragmatist, rebuffed her critics. Working with industry was essential to raise recycling rates, she said. In a June 2004 editorial for *Plastics News*, Krebs said BPEC members were more focused than ever on increasing recycling. 'I expect great things will come from the work of BPEC,' she wrote.

Facilitating the talks was a good money-maker for NRC. Each participant paid the organisation $10,000. The drinks makers also covered half the salary of a new director of policy and programmes at NRC, and took care of any additional expenses. For the drinks companies, conducting research under the oversight of a respected recycling body lent them credibility as they searched for ways to inexpensively boost recycling rates, fend off restrictive regulation and avoid bottle bills. The partnership promised to be symbiotic.

Krebs says her aim in bringing more corporations into NRC's fold was to create a forum for companies and grassroots recyclers to discuss strategies and solutions. 'There had not yet been a table that environmental activists could sit at with the corporate folks like a Scott Vitters,' she says.

In the case of BPEC, however, there was no such forum. The research process was kept highly secretive and nobody inside NRC, other than Krebs, was privy to it. BPEC's research was spearheaded by Kevin Dietly at Northbridge Environmental, the same consultant the soft drink trade association had hired to denounce the BEAR report. Dietly had also done work for the plastics industry back in the early 1990s to fight Steven Englebright's ban.

A sharp dresser who favoured crisp white shirts and sleekly cut suits, Dietly wore rimless glasses and spoke in a measured – some might say slightly wooden – voice. He had testified against bottle bills on behalf of the industry since the 1980s, always arriving at hearings impeccably prepared and laden with facts and figures he had memorised. Krebs, who as a supporter of California's bottle bill had once been in an opposing camp to Dietly, was now working shoulder to shoulder with him.

Dietly had spent decades arguing that bottle bills were a tax on consumers, that they did nothing to cut litter and that they unfairly targeted bottles and cans, which were a small slice of what was thrown away. He also said they harmed overall recycling by removing the most valuable recyclables from kerbside recycling programmes, making these economically unviable.

The environmentalists contended that using bottles and cans to subsidise everything else didn't solve the underlying problem – the panoply of plastics that were uneconomical to recycle remained money losers. If anything, those plastics should be banned or taxed to pay for their own disposal. They also pointed out that many kerb-side recycling programmes had for years been unprofitable, even in states that didn't have bottle bills. Drinks bottles were widely littered and also one of the most commonly discarded consumer goods items in America. The environmentalists also said kerbside recycling programmes didn't capture the large slice of soda and bottled water drunk on the go. This last argument was popular – and one the drinks industry had long wrestled to combat.

At bottle-bill hearings, Dietly often pointed to the declining rates at which people returned containers in many deposit states. The drops, he said, indicated that people found returning containers inconvenient and that bottle bills didn't work. What Dietly omitted to explain was that industry lobbying had ensured the deposit stayed at 5 cents in many places rather than rising with inflation, inevitably lessening people's motivation to go to the trouble of getting their money back. Among his other arguments were that introducing bottle bills would raise costs for consumers and eliminate thousands of jobs. Year after year, elected officials hearing these arguments were scared off and bottle bill proposals fizzled.

But by 2003, recycling rates across the country were continuing to flounder and there were clear signs that the drinks industry's arguments weren't holding as much weight. Despite ferocious industry opposition, in 2002 Hawaii had signed a bottle bill into law. It was the first state to do so since 1986. A prominent senator from Vermont, Jim Jeffords, was proposing a national bottle bill that would require soda, beer and other drinks makers to charge a 10-cent refundable deposit to raise recycling rates.

A bottle bill battle was also playing out on the streets and in the bodegas of New York. Environmentalists wanted to expand the state's bottle bill to include bottled water, fruit juices and sports drinks; these hadn't really existed when the law was enacted in 1982. They also wanted the unredeemed deposits – so far left with companies – to be handed over to the state, which would use these to improve recycling programmes.

Under Goizueta – the CEO who preceded Ivester – Coke had doubled down on bottled water. The company saw any regulation that might slow sales of its recently launched water brand, Dasani, as a major threat. Pepsi, which sold its own bottled water brand, Aquafina, had the same concerns. By the early 2000s, bottled water had become crucial for the drinks makers who were seeing a slide in soda consumption. In 2001, bottled water sales grew 30 per cent while fizzy soft drinks rose just 0.6 per cent.

And so Coke and Pepsi formed a lobby group. Its name – New Yorkers for Real Recycling Reform – offered little indication about its corporate backers. The group, which also represented New York grocers, created thousands of anti-bottle bill posters for retailers to prominently display. One poster featured a giant cockroach. 'He supports the bigger bottle bill!' it proclaimed, warning of 'major mess and hassle for consumers'. Environmentalists drily pointed out that the deposits were already being charged on sugary sodas and beer, and that the expansion would mainly target bottled water, which wasn't known to be a huge draw for roaches.

Coke and Pepsi's loudest opponent was a vivacious young woman named Laura Haight, who had an infectious guffaw and a dishevelled mane of long dark hair she wore loose. Haight worked for the New York Public Interest Research Group, or NYPIRG. She would spend seven years fighting for the expansion of the state's bottle bill, a campaign she named 'The Bigger, Better Bottle Bill'.

While Coke and Pepsi had millions of dollars and some of the country's most expensive lobbyists at their beck and call, Haight had a 20-foot inflatable Snapple bottle and an army of student volunteers from 20 universities around the state. The students collected data on what was being littered, wrote letters to legislators and showed up in force at press conferences and rallies.

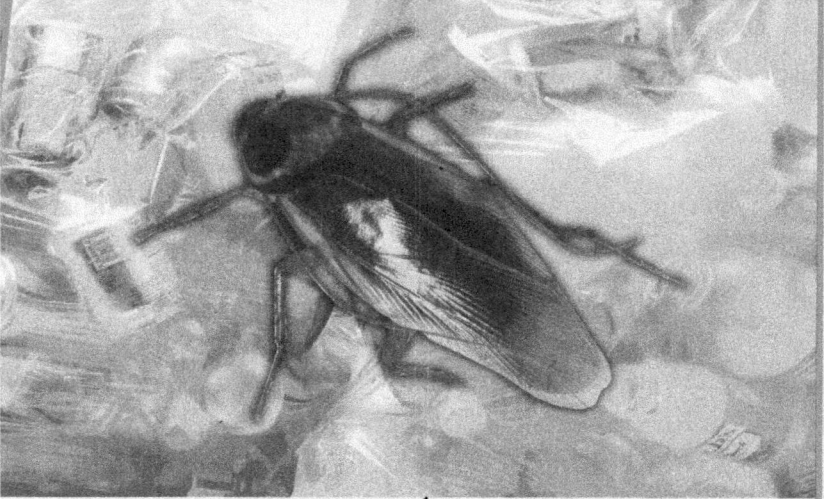

He supports the Bigger Bottle Bill.

Expanding the current bottle bill to include 5-cent deposits on nearly all beverage containers including juice, tea, bottled water and sports drinks will create a major hassle and mess for consumers.

New Yorkers will have to store more containers in their homes, haul more dirty containers back to the store and endure longer waits to recycle.

Instead, New York should expand its current convenient and economical recycling programs, making recycling easier, cleaner and more convenient for all New Yorkers.

And that's bad news for any unwanted guests.

Tell the Albany politicians not to raise the cost of groceries while making life more difficult for everyone except the roaches.

New Yorkers for Real Recycling Reform
Clean and simple.

Visit www.realrecyclingreform.com for more information.

Image courtesy of Laura Haight

One weekend, Haight got 2,300 students to hold hands and ring New York's capitol building, forming a human chain to represent the number of bottles thrown away in the state every 30 seconds. Haight also had them dance in parks and on street corners while wearing life-sized photocopies of a nickel mounted on cork board to raise awareness about the need to expand the 5 cent deposits. 'My dancing nickels were very easy to dance in,' she says. 'I didn't force anyone to dance but it was recommended.'

On St Patrick's Day, students dressed up as leprechauns stood next to buckets filled with gold-painted bottle caps to advertise the 'pot of gold' New York was missing out on by leaving the unclaimed deposits with the soft drink industry. 'We got a lot of news coverage from these silly things,' says Haight. 'We had very little money but were very creative.'

The drinks makers were not amused. Much like the fast-food industry and diaper makers had before them, they argued that drinks containers were a tiny slice of America's waste problems and were being unfairly singled out. 'Beverages was the whipping boy for the issues,' Dietly says. 'It was the default that we needed to do something.'

Against this backdrop, in August 2005, Dietly took to the stage in front of hundreds of attendees at NRC's flagship annual conference to present BPEC's findings. He explained that while data on sales and recycling had previously been available through industry associations, the companies now had data on *where* their drinks were consumed, attained through consumer behaviour studies.

BPEC's main finding was that, contrary to common belief, the majority of drinks were consumed in people's homes rather than on the go. The drinks companies' implication was clear: the focus needed to shift to developing kerbside recycling programmes (paid for by governments) and away from bottle bills (paid for by the drinks industry's consumers). 'We realise that home is still the place where we have to keep our efforts,' Dietly told the audience. 'It would be unfortunate to continue to move forward on the basis of anecdotal information.'

The report, for which NRC paid Dietly's firm Northbridge $140,000, ruffled some feathers. The Container Recycling

Institute – a prominent NRC member – demanded disclosures about the financial relationship between NRC and BPEC. It sought a copy of the consumption data and asked why viewpoints beyond those from the beverage industry hadn't been incorporated. BPEC said it wouldn't share the data, citing competition concerns. 'In addition to the questions CRI still has about the BPEC research, we are concerned about the lack of transparency,' wrote the non-profit's head Pat Franklin in CRI's 2005 newsletter.

Franklin – perhaps unsurprisingly, given her advocacy for bottle bills – questioned whether the industry's numbers were accurate. Even if they were, she maintained, it was clear that kerbside recycling had failed to raise container recycling rates. She pointed out that although the number of Americans who had kerbside recycling programmes available to them had grown 79 per cent to 140 million in the decade to 2001, the nation's recycling rates had fallen dramatically over this period. The recycling rate for PET bottles had plummeted from 38 per cent to 22 per cent, while the rate for aluminium cans had dropped from 65 per cent to 49 per cent. 'Do we really want to pin our hopes on curbside recycling?' Franklin wrote in an op-ed published in *Waste & Recycling News*.

BPEC's response was that the kerbside data was poor and didn't reflect reality. Touting NRC's role as the facilitating body, the drinks industry's report grabbed headlines and gave new ammunition to its fight against bottle bills. Kate Krebs herself hailed the results as 'groundbreaking' and defended the drinks makers against allegations of bias. She told a reporter: 'NRC's credibility is utmost and we wouldn't be involved with something that didn't have a lot of honesty and integrity'.

*

In March 2007, a letter landed on the desk of NRC's president, David Refkin. The letter was from Richard F. Roy, chairman of the Connecticut Environment Committee, who was sponsoring SB 1289, a bill to expand the state's container deposit law to include bottled water.

'On March 7th Ms. Krebs was in the state capitol accompanied by a group of industry lobbyists who are opposed to SB-1289,' wrote Roy. Krebs, he added, had spent much of the day meeting with house and senate members and had tried to convince him that a single-stream kerbside recycling programme – in which all recyclables go into one bin that's collected by a waste hauler paid for by the municipality – was a better system than a bottle bill.

'I gathered from Ms. Krebs' negative comments about bottle bills and the timing of her visit (a crucial time for the expansion bill) that your organisation was on record as opposing bottle bills,' wrote Roy. 'Absent such a position, it seems oddly coincidental that the executive director, representing NRC's members, would be here in Connecticut with lobbyists working against my efforts to increase recycling.'

SB 1289 did not pass that year.

Inside NRC, news of Krebs's stealthy lobbying sparked an uproar. For many, the letter confirmed the suspicions they had harboured for months. Krebs – the once small-town recycler who had been an ardent proponent of California's bottle bill – was now firmly aligned with Coke and other drinks makers in opposing container deposits. 'The Connecticut Recycling Association was livid,' recalls Refkin. 'This became a big deal at NRC.'[3]

---

[3] When I first asked Krebs about her opposition to the expansion of the Connecticut bottle bill, she told me she had only opposed the state keeping unclaimed deposits since these went to a general fund and weren't used to fund recycling. After I read her the letter written by Robert Roy detailing her opposition to the bottle bill's

By now, Coke had not one but two executives on NRC's board – Scott Vitters had been joined by Gary Wygant, a mouthy erstwhile aluminium industry executive who now worked for the company's bottler, Coca-Cola Refreshments. Such strong representation by one brand was unheard of in NRC's history.

Krebs had also taken a month away from the recycling body to visit Antarctica on an 'eco-tour' courtesy of Coke. The soft drink giant hosted the all-expenses-paid trip for a select group of employees and outsiders to visit the continent. Krebs flew to Buenos Aires, did several days of team-building exercises and then moved on to Ushuaia, the southernmost city in the world.

The trip didn't sit well with many of the recyclers who Krebs had spent decades working alongside. 'It really made people mad,' says Jerry Powell, a one-time NRC board chair. 'Boy, if you want to talk about cosy, that is about as cosy as it gets, taking such a

---

expansion, in which Roy says Krebs told him that bottle bills and kerbside recycling compete and don't work well together, Krebs initially stuck to her original position. She then said she opposed the Connecticut bottle's bill expansion because it would be taking valuable PET and aluminium away from kerbside programmes, denying municipal recyclers revenue and making their programmes unviable. I pointed out that there are many ongoing kerbside recycling programmes that run successfully in states that have bottle bills, and there are states in which kerbside programmes – despite the absence of bottle bills – have struggled so much that they've closed down. Krebs denied she had told Roy that bottle bills and kerbside programmes couldn't coexist. After my pressing further on why the letter would say this, she said: 'What I'm also saying is that I support kerbside recycling, I support single-stream recycling because consumers find that convenient. Period.' In a separate exchange with me Krebs said she was in the Connecticut capitol mainly to get legislators to support a recycling programme the drinks companies were funding in Hartford. She said that while I was trying to paint her as having changed while at NRC from pro-bottle bills to against them, in fact she had long been in favour of single-stream recycling and had only supported California's bill because unclaimed deposits flow to recyclers rather than to the state.

major gift.' Krebs says the trip was to learn about climate change and she sought the board's permission beforehand. 'We were on a Russian icebreaker research boat and it was amazing,' she recalls. 'I have no idea of the cost.'

By 2007, Krebs was being compensated more than $140,000 a year, up from $90,000 when she started as executive director. NRC that year reported $1.2 million from grants, gifts and contributions, up 178 per cent from what it made in 2002 before Vitters joined the board.

'As the money pendulum swung a bit, Kate swung with it. She migrated towards where money was coming from,' remembers George Dreckmann, who was NRC's chairman at the time and would often bring a bottle of Pepsi into board meetings to thumb his nose at the Coke camp. 'It didn't feel right, but it wasn't criminal,' he says. 'I just started to feel like I should take a shower.'

NRC's board with Kate Krebs pictured standing on the left in the second row. Image courtesy of David Refkin

For her part, Krebs maintains that meeting the high expectations of the local recycling coordinators who were NRC's main members required money.

As NRC had taken more corporations on board and received more funding, Krebs was also spending increasingly large amounts. In 2006, she had moved NRC into a swanky new office that had a marble-floored lobby and individual offices for all staff members in a sought-after part of downtown DC. The rent was $163,254 for the first year, with subsequent increases baked into the lease agreement. Staffers and board members wondered aloud whether the offices weren't far too tony for a recycling non-profit. 'People liked to say that NRC was on a beer pocketbook and living a Dom Perignon lifestyle,' remembers Mark Lichtenstein.

By now, Krebs's relationship with key members of NRC's board was souring. In a May 2007 letter to Krebs, Refkin, Dreckmann and the board's vice-president Susan Kattchee wrote that they had 'serious concerns' about Krebs's leadership. They complained that Krebs was increasingly going her own way, making hires and investment decisions without consulting them. 'There seems to be a basic lack of understanding or acceptance that you report to the Board of Directors,' they wrote.

The three board members also wrote that the time had come for NRC to set up both a whistleblower policy and a conflict of interest policy. 'As you know the "left flank" of the industry has concerns about "industry influence",' they warned. As her relationship with Coke had deepened, Krebs had begun making regular appearances in news articles and press releases in which she praised the company for its recycling efforts.

In 2007, Coke announced that it would open the world's largest PET bottle-to-bottle recycling plant in Spartanburg, South Carolina. Without offering any timeframe, Coke also declared an aim to recycle or reuse 100 per cent of the plastic bottles it used in the US. It would set up a new entity, Coca-Cola Recycling, to fulfil this aim. The company reminded people that it had been 'focused on' PET recycling since it introduced the first drinks bottle made with recycled material in 1991. It said Coke used recycled content in more than 17 countries around the world, including the US. Notably, it was silent on what percentage of its bottles it did this for, or even what percentage of recycled content it used.

The press release announcing Coke's recycling plant and new targets included this quote from Krebs: 'Coca-Cola has staked a clear leadership position in its approach to sustainable packaging. The new Spartanburg plant represents an end-to-end recycling model that is world class and that I hope other industries will follow.' A newspaper article about the Coke plant in the *Atlanta Journal-Constitution* contained another comment from Krebs: 'That kind of commitment is head and shoulders above other companies throughout the world.'

By now, the latest available data for PET drinks containers showed a US recycling rate of 23 per cent, a full 16 percentage points down from the level a decade earlier in 1996.

Coke's various promises nonetheless attracted a deluge of positive publicity. In April 2008, *Fortune* magazine ran a profile of Coke's CEO Neville Isdell with the headline 'Coca-Cola's Green Crusader'. The article included a photo of Isdell – incongruously wearing an expensive grey suit as he sat on a wooden chair in the middle of a sunny field – cheering with a bottle of Coke. The bottle was made of glass, despite the fact that the material by now made up only 1.2 per cent of the US soft drink market. 'It helps when the CEO is committed to sustainability' was the subtitle that followed. The reporter went on to describe Coke's sustainability efforts, including the recycling plant. 'They are willing to try as many avenues as they can,' was the approving quote from Krebs, whom the *Fortune* reporter identified simply as being 'of the National Recycling Coalition', seemingly unaware of Coke's connections with NRC.

Over the previous two years, Coke and its trade bodies had sponsored a number of initiatives that NRC had been involved in. These included campaigns to 'rebrand recycling' by educating consumers and simplifying recycling labels, along with a scheme to offer gift cards and vouchers to residents in 'model cities' like Hartford, Connecticut, if they recycled properly. Coke individually was contributing more than $100,000 a year to NRC. The BPEC project brought in a further $210,000.

Coke's hope, of course, was to improve recycling rates. But it also wanted to keep mandatory container deposit laws at bay by showing regulators how recycling was best done through kerbside programmes for which taxpayers picked up the tab. 'All of these beverage container makers were fearful of policy solutions with exorbitant costs. Clearly it was in their best interests for costs to be minimised,' says Ed Skernolis, an NRC staffer hired by Krebs who ran the model cities programme. 'One of the ways to play defence on that is to see if you can develop grassroots campaigns that enhance recycling at the local level.' To Skernolis, there was an inherent conflict between the bulk of NRC's membership and the companies pumping it with money. 'The membership largely wanted policy-based solutions and the people giving it money wanted market-based solutions.'

Particularly as public opinion against bottled water began to sour, the industry highlighted the work it was doing with NRC to argue against regulation that threatened to cap bottled water profits. Coke's deepening partnership with NRC – and Krebs's many idolising public comments about the drinks giant, made from the vantage point of a seemingly neutral recycling industry expert – undoubtedly burnished Coke's reputation.

Only two months before the *Fortune* article ran, NRC had named Coke the winner of its coveted 'Recycling Works!' award, citing the company's recycling investments and its goal of recycling or reusing 100 per cent of its plastic bottles in the US. 'The Coca-Cola Company has played a pivotal and unprecedented leadership role with their commitment to a sustainable economy and a sustainable environment,' said Krebs in the widely disseminated awards announcement. 'We enthusiastically applaud their leadership and honour their commitment.'

In effect, Coke's harnessing of Krebs was a tactic similar to the one Shelby Yastrow had used in partnering with the Environmental Defense Fund to launch a waste-reduction campaign to rebuild McDonald's reputation in 1989: why say something nice about yourself if you can get someone credible and widely respected to say it for you?

'Some would say that Coke infiltrated NRC,' says Jerry Powell, the former board chair. 'And I agree with that.'

*

On a cool, dry Wednesday in January 2009, Coca-Cola's president of North America, Sandy Douglas, threw open the doors to the company's shiny new plastics recycling plant on North Blackstock Road in Spartanburg. Douglas had worked at Coke for more than two decades, climbing his way up after starting out as a regional sales manager. At his core, he had stayed a salesman, and selling Coca-Cola's newly invigorated vision of itself as a responsible corporate citizen was a role that came naturally to him.

Douglas looked self-assured as he stood before the state officials, recyclers, journalists and others gathered for the opening. Coke's new plant, he said, would be able to produce the equivalent of 2 billion 20-ounce plastic bottles a year. This would prevent 1 million metric tons of carbon dioxide from entering the atmosphere in the following decade, the equivalent of eliminating 215,000 cars from the road. Coke had spent $44 million on the plant and would invest a further $16 million on increasing kerbside recycling through education and incentives, added Douglas.

Notably, Coca-Cola's plan was to source virtually all the used plastic for the plant from kerbside bins rather than container deposit programmes. In 2009, both New York and Connecticut successfully expanded their bottle bills to include bottled water, piling pressure on Coke to move quickly to forestall similar regulation elsewhere. If Spartanburg was successful, Coke hoped it could show America once and for all how container deposit programmes were unnecessary.

Gazing benevolently upon his assembled guests, Douglas asked them to raise a toast with bottles of Coke Classic made especially for the event from 30 per cent recycled plastic. 'We believe environmental stewardship is more than philanthropy,' he said. 'This Spartanburg plant will be a model for sustainable green business in the years to come.'

Coke had spent months planning the PR for the plant's opening. 'Those sorts of plant openings, there's not a lot of glamour to them,' says Lisa Manley, who worked as the company's head of sustainability communications at the time. 'This plant was not in the most media-friendly location and we wanted to get the news out while knowing not a lot of folks would come to the facility in Spartanburg.'

Key to Coke's PR effort was Kate Krebs.

Krebs had resigned from NRC a few months earlier in July 2008 after a serious disagreement with its board members over the organisation's funds. These were divided into restricted pots (donated by companies for specific programmes) and unrestricted ones (for instance, from NRC's membership fees and its annual conference).

The expensive lease on the new office – which NRC president David Refkin says Krebs signed without seeking approval – turned out to be the tip of the iceberg. The organisation's expenses rose

84 per cent in the two years to March 2008, while revenue climbed just 24 per cent, leaving NRC in the red.[4] 'Our financial house must be brought into order,' Refkin had written in a letter to Krebs in April 2008. 'We need to seriously evaluate our cash position and overall financial health.'

Soon after, Krebs allegedly used restricted funds designated for the model cities and other programmes to pay general expenses like salaries. 'Kate was managing by taking from one pocket to put in another,' explains Gary Wygant, the Coca-Cola executive who served on NRC's board alongside Vitters. This same allegation was made by five other NRC board members and staffers I interviewed who worked with Krebs at the time.

NRC's auditor hadn't flagged any irregularities and the finding came as a shock to the board. George Dreckmann, David Refkin and NRC treasurer Stampp Corbin all say they tried to bring in a new auditor to look at NRC's accounts but Krebs fiercely resisted. Eventually, she hashed out a resignation deal with the board, non-disclosure agreements were signed all round, and a suitable press release was published, thanking Krebs for her service. By July 2008, she was gone.[5]

---

[4] Krebs says her spending on the new office reflected a grant she expected NRC to get that never materialised.

[5] Krebs says she left with a clean audit. When I asked if she did tap restricted funds, she said: 'I don't think I did.' She added: 'The audit would have caught if there were things that were not done correctly.' George Dreckmann, NRC's chairman, says the audit didn't properly reflect what was going on with the organisation's financials and questions why Krebs was so against allowing a new auditor to review the financials. 'We felt the company that had done our previous audits had overlooked the practices which concerned us and would not provide an objective review,' Dreckmann told me. 'The fact that she was so opposed to this, and our subsequent financial issues, indicated to me that what she was doing was indeed in violation of NRC policy and that she had to be removed.'

Ed Skernolis took over as interim executive director, sharing the role with another of Krebs's hires. 'At first, I didn't realise how bad it was because Kate had kept the finances under wraps,' says Skernolis. Once he reviewed NRC's cash flow and debts, he wrote a memo to the board. 'We are in trouble,' it said. 'We need significant new revenues to sustain this organisation.'

Krebs, meanwhile, had moved on to another organisation: the Climate Group. From her new perch, she soon made another appearance on behalf of Coca-Cola to drum up publicity for the Spartanburg plant. On 15th January, 2009, Coke launched a carefully orchestrated media blitz, lining up dozens of back-to-back interviews with Scott Vitters. He was joined by Krebs, who explained to Americans how and why they should recycle, while also heaping praise on Coke for its efforts. Reporters who quoted Krebs described her as working for 'a leading recycling advocacy organisation'.

Not mentioned in Coke's announcements that day was that the Climate Group counted the drinks giant among its corporate members. The Climate Group had also been working with the American Beverage Association to emphasise that the soft drink industry had done its part by making their products recyclable – now the onus was on consumers and local governments to do theirs. 'To see Coca-Cola take the issue of climate change head-on is wonderful,' Krebs told a reporter for the *Herald-Journal* in Spartanburg. Suitably impressed, the reporter went on to conclude: 'Coca-Cola's efforts met the expectations of at least one environmental watchdog group that has encouraged the company to take braver actions.'

Other news outlets across the country also carried admiring stories about the company's dedication to sustainability. Coca-Cola's media relations department was named a finalist at *PR News*'s annual Corporate Social Responsibility Awards. The company also won a gold medal award from the World Environment Center for its work on 'sustainable packaging'.

The wave of good publicity was only amplified in 2009 when Coke's CEO Muhtar Kent announced that the company was standing on the precipice of a major breakthrough. Coke, disclosed Kent while sitting on stage at an event alongside former US president Bill Clinton, would soon begin making its plastic PET bottles partly from sugarcane-based plastic that could also be recycled. 'That single initiative will transform the whole concept of recycling around the world for beverage containers,' he said.

In the months that followed, Coke made sure everyone knew about the plant bottle. Samples were readily available at its investor events in Atlanta, at the UN's climate change summit in Copenhagen and at the Winter Olympics in Vancouver. Krebs made another appearance in a Coke press release to applaud the company. 'The PlantBottle is precisely the kind of innovation that demonstrates how businesses can help address climate change and reduce stress on our precious natural resources,' she said. 'This is a revolutionary solution that has the potential for long-term, meaningful benefits.'

Coke vowed that, by 2020, *all* its plastic bottles around the world would use up to 30 per cent plant packaging. Scott Vitters went a step further and announced that Coke's goal was 'to make traditional plastic bottles a thing of the past' and to ensure that every drink Coke produced was available in bottles made entirely from plant-based plastics. In 2011, *Fortune* named Vitters one of its 'eight green stars' for his work on the plant bottle.

Riding on the coattails of all the goodwill it was amassing, Coke also kicked off a multi-million-dollar marketing campaign called 'Give it Back', which featured the new Spartanburg recycling plant. The debut ad ran during Fox's popular show, *American Idol*. In it, bottles and cans placed in recycling bins subsequently popped out of vending machines or showed up on supermarket shelves as good as new. Coke's research showed that more than 30 million people were tuned in when the ad played. The company said its aim was to ensure consumers saw versions of the ad across TV, print and online a billion times.

'If you've had a Coke in the last 40 years,' cooed the ad's voice-over, 'you've played a part in one of the largest beverage recycling efforts in the world.'

*

By 2011, things at the Spartanburg plant had fallen apart.

Just as McDonald's had been waylaid in its attempts to recycle its clamshells, Coke had discovered that bales of what should have been PET plastic arriving on trucks were so contaminated that sometimes half a load would need to be landfilled.

'We took this hideous material because we were ideologically opposed to bottle bills and the result was impurities embedded in the plastic which led to lack of quality,' says Jeff Seabright, who worked as Coca-Cola's vice president of environment and water resources for more than a decade until 2014. 'The end product had a failure rate of one in 100 and Coke's standard was something like one failure in 100,000, so it was effort after effort after effort.'

While working for Coke, Seabright publicly denounced bottle bills, saying they didn't raise overall recycling rates. But, looking back, he is candid about the company's motive: to keep costs as low as possible. 'Our opposition was rooted in naked self-interest not in a higher purpose, despite the fact that we positioned it that way.'

Coke had zeroed in on a Spartanburg recycling firm called United Resource Recovery Corporation to run the plant. URRC had what seemed like an innovative new technology that meant it didn't need to melt used plastic to remove contaminants, a practice that commonly discoloured plastic, making it unsuitable for new bottles. 'You melt crap and it just gets dispersed into the bottle, which may turn yellow or grey,' explains Carlos Gutierrez, who headed URRC. Instead, URRC used a diluted sodium hydroxide solution to remove oils and other contaminants. The plant also had what it promised was cutting-edge technology to separate out PVC labels which, even in small quantities, turn PET brittle and yellow.

But things didn't work out as planned. With every sort, there was less plastic left to recycle, driving up costs. The plant was burning through $1.7 million in cash each month. Making a bad situation worse, the price of virgin PET dropped alongside oil prices. Coke's bottlers refused to buy the recycled plastic, which was both lower-quality and more expensive. 'The bottlers are chasing pennies down the corridor every day,' says Seabright. 'We couldn't generate anything usable coming out of that plant. We were selling at a massive discount just to offload it.'

By relying only on kerbside programmes, the sorting facilities that supplied Coke had to pull out plastic that was mixed in with paper, glass and aluminium, and then further sort this into PET bottles. Single-stream recycling saves municipalities money on collection since they need to provide residents with one large bin rather than several, and recycling trucks can dump everything into a single compartment. But the costs show up in other ways. Contamination rates from non-recyclables – as well as between the various headline categories of recyclables – are much higher, making the sorting process far more tedious and expensive.

Coke was also facing a shortage of used PET bottles. More than half of those collected in the US were exported to China, where sorting was far cheaper.

By 2011, Gutierrez was forced to suspend operations at Spartanburg and lay off half his staff. Coke brought in another company, International Recycling Group, to run the plant, but it too ran into similar problems. In 2014, the plant, very quietly, closed its doors for the final time.

By now, it was 23 years since Coke had first tested a plastic bottle that incorporated recycled material. The percentage of PET bottles being recycled into new food or drink containers stood at just 6 per cent.

The National Recycling Council, as it was first conceived, was also a thing of the past.

Fourteen months after Krebs left NRC, the organisation had voted to file for bankruptcy and laid off its staff. Its debt stood at $1.6 million, a chunk of which was tied to the lease Krebs had signed on the swanky DC office. NRC had also been forced to cancel its 2009 annual congress after most recyclers said they wouldn't attend. The financial crisis had torpedoed their budgets, but many long-time recyclers had also lost trust in NRC. One described the organisation as 'the best corporate recycling association in America, because that's what it's become . . . it shouldn't pretend that it's the voice for communities and municipalities.'

The cancellation of the congress left NRC with just $619 remaining in its bank account. Coke took its money to Keep America Beautiful – the creator of the Crying Indian ad – which had long emphasised that the responsibility for recycling lay with the individual. Eventually, NRC found a way to break its expensive office lease by forfeiting the $12,000 security deposit, allowing it to narrowly avoid bankruptcy. But former NRC members say the organisation never recovered and, even today, is a shell of its former self.

As for Krebs, in 2013 she landed a new job working directly for the American Beverage Association. As its senior environmental policy advisor, she far more openly lobbied against bottle bills. 'Recycling is important, but what matters most is how we recycle,' she told a journalist in 2013. 'Container deposits lead to higher costs, more economic dislocation and unintended consequences, regardless of how the programs are designed.'

No container deposit law has passed in the US since Hawaii's in 2002. Today, more than five decades after Coke and Pepsi began using plastic bottles, the latest available data shows that the US's PET container recycling rate stands at 29 per cent.[6]

---

[6] By contrast, Germany's PET bottle recycling rate stands at 95 per cent, while Denmark's is 96 per cent. Both countries have well-functioning deposit programmes: https://www.unesda.eu/pet-collection-rates/

PART IV

THE NEW PLASTICS BACKLASH

CHAPTER 12

THE LAST STRAW

'A problem that gets bigger the smaller it becomes'

The sun was high in the sky when Christine Figgener pulled a turtle onto the rented motorboat she was idling some 12 miles off the Pacific coast of Costa Rica.

The lanky marine biologist from the German town of Marl had been out on the water with a seven-person team since 5.30am to study turtles. It was after lunch and her energy was starting to flag. Still, Figgener carefully measured the male Olive ridley, took a small tissue sample to use in a genetic study and attached flipper tags. Ordinarily she'd have then plunked him back in the water, but a fellow marine biologist noticed that the turtle had what appeared to be a barnacle stuck up his nose.

'I'm going to get that out,' he told Figgener, snapping open the pliers on a Swiss Army knife.

Having little else to do, Figgener – whose fascination with the ocean had started at age two on a family vacation to Greece – began filming the extraction. 'He's going to be so happy!' she exclaimed. 'Big booger.'

'You know what this is?' said the biologist, Nathan Robinson, after trying unsuccessfully to get a grip on the object. 'That's a worm.'

Figgener shuddered. 'Oh, that is disgusting,' she said.

The turtle was really writhing by this point. Blood had begun trickling out of his nose and he was clearly in pain. Eventually, Robinson gave up, instead cutting the obstruction, which by then was sticking an inch out of the turtle's nose.

'What is it?' asked Figgener, leaning over, camera still in hand.

'Plástico,' offered a Costa Rican crew member. 'Plástico!' repeated Figgener. 'Don't tell me it's a frickin' straw!' she exclaimed. 'This is the reason why we do not need plastic straws!'

Turtles feed underwater, in the process swallowing sea water they expel through their noses. This particular turtle would have tried to also expel the straw, which is a risky business, Figgener tells me, given how narrow the nasal cavity is. She says the straw was so difficult to remove because scar tissue had likely formed, which indicated the offending object had been in the turtle's nose for several weeks.

Once he realised what it was, Robinson gave the extraction another go. He kept yanking at the remainder of the straw but the pliers kept slipping. The turtle struggled, hissing repeatedly in fear, but was powerless, restrained by crew members.

Eventually, after nearly eight minutes of trying, the straw gave way, sliding out.

Back at the house she was renting in Playas del Coco, Figgener put the video she had shot on YouTube, leaving it to upload over several hours using the weak Wi-Fi signal from the neighbouring pizza restaurant.

As I write this in the summer of 2024, Figgener's video – shot on the afternoon of 10th August, 2015 – has been watched more than 110 million times on YouTube. Versions posted elsewhere have been watched millions more times. You can find it online if you're curious, but it makes for hard watching.

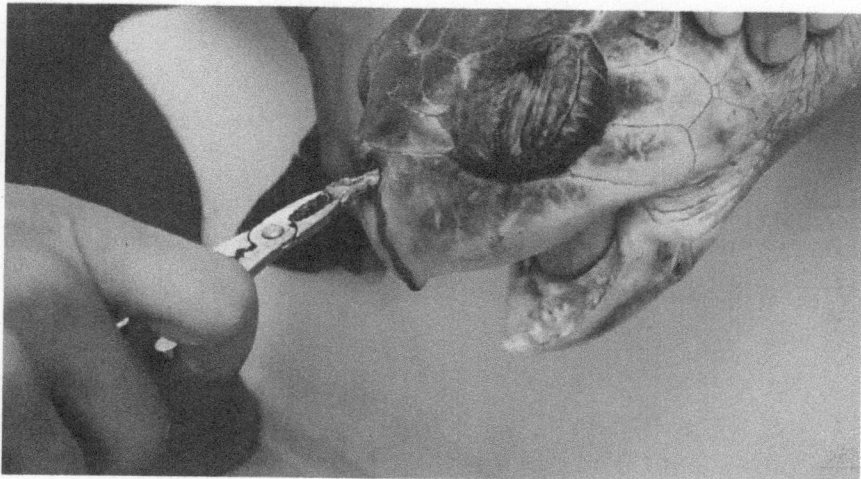

A still from Figgener's video showing the straw being extracted from the turtle's nose. Image courtesy of Christine Figgener

For decades before Figgener found the Costa Rican turtle, researchers had been documenting the impact of plastic waste in the oceans. Study after study detailed how plastic was entangling animals or causing intestinal blockages that left them unable to eat and ruptured their internal organs.

In 1985, at the first international conference on marine plastic waste, held at the University of Hawaii, researchers announced that between one and two million albatrosses, puffins, gulls and other sea birds were dying each year from entanglement in plastic fishing nets or from eating particles of plastic. Another 100,000 sea mammals – such as whales, dolphins, seals and manatees – were meeting similar fates.

Hundreds if not thousands of stories had been published about the effect plastic was having on sea life and biodiversity. But most were before the era of social media. Figgener's video was the one that really stopped people in their tracks.

'It's almost like a bloody car accident that is really horrible, that nobody wants to see but nobody can not watch,' says Figgener. 'The turtle all of a sudden gave those statistics published in research articles an actual face.'

*

Nearly three decades after the *Mobro*'s voyage sparked an American awakening about waste, the turtle propelled a new backlash against plastics.

In an era of social media, this wave of anti-plastic sentiment crashed across national borders to go global. It metastasised from concerns about the impact of visible pieces of plastic on ocean life to also encompass the more insidious threats of microplastics and the thousands of potentially hazardous chemicals found in plastics.

In the months that followed, a rising number of towns, cities and states moved to ban or restrict plastic straws,[1] cutlery and bags. Over the next five years, the number of countries implementing restrictions on single-use plastic more than doubled. The targeted products individually made up only small slices of the overall waste stream, but they accounted for a substantial amount of the litter found on beaches and in the oceans. Their visibility sparked a global conversation about the mismatch between using a material that lasts hundreds of years to make products designed to be used for a few seconds.

---

[1] Plastic straw bans inevitably lead to paper straws being used as replacements. Apart from their functional limitations, paper straws aren't much greener than plastic. Paper straws are rarely recycled because they're too small to be sorted from mixed recycling streams. Even if they're put in paper-only recycling bins, paper mills say they don't want them as the straws often contain residue from smoothies or milkshakes which can weaken the fibre and contaminate water systems. Paper straws also use more material than plastic ones and can generate higher emissions. They've been found to contain PFAs. That being said, the main reason governments banned plastic straws is litter, which can directly impact living creatures. Concerns about emissions and recycling pale in comparison.

Between the industry's inception and 2015, around 6,300 million metric tons of plastic waste had been generated. Only around 9 per cent of that had been recycled. A whopping 79 per cent had gone to landfills or leaked into nature.

In 2016, Google news searches for plastic began to spike in the US, India, France and other countries. Consumer goods companies noticed a gradual shift in behaviour as people looked for ways to pull back on plastic. By now, the public was also hearing about a far smaller kind of plastic litter found mingled among sand on beaches, floating in oceans, and in the bodies of sea animals. The presence of small plastic particles in the oceans had been reported since the 1970s, but it wasn't until the early 2000s that the word 'microplastics' started popping up. I first found it used in a March 2002 *Los Angeles Times* story. The term is commonly used today to refer to particles ranging in size from 0.1 micrometres to 5,000 micrometres.

In 2004, Richard Thompson, a marine biology professor at the University of Plymouth on the south-west coast of England, published a paper called 'Lost at Sea'. In it he described how microscopic plastic fragments and fibres, resulting from the degradation of larger plastic items, were 'widespread' in the oceans and being ingested by marine organisms. His paper, which ran in the academic journal *Science* over a long weekend in May, made a splash. When Thompson arrived at his office on Tuesday morning after a restorative three days spent camping, he had dozens of emails and voicemails from journalists.

In 2007, Thompson wrote a second paper on microplastics. This one revealed that the plastic bits were carriers for hazardous chemicals. His study showed that microplastics soaked up phenanthrene – a common contaminant produced by burning carbon-based materials – to a far greater extent than sand did. It also showed that in marine worms exposed to such microplastics, the concentration of phenanthrene rose sharply.

It was becoming apparent that not only were microplastics veritable sponges for all kinds of toxic chemicals, but that they were also easily mistaken for food by tiny animals who then ingested the chemicals along with the plastic. A science writer covering Thompson's study at the time described microplastics as 'tiny Trojan horses'.

Many of those microplastics were coming from the abrasion of car tyres and the washing of synthetic clothing. Run-off from roads swept microplastics down drains and into the oceans, while microfibres from washing machines persisted in sewage sludge, which could then end up in the soil as fertiliser. But tiny plastic particles were also being intentionally added by consumer goods companies to everyday products like facial scrubs and toothpaste. Called 'microbeads', the plastic bits scrubbed off dead skin and stains from teeth, before washing down the drain and into the ocean.

Eating chemicals, it turns out, isn't good for sea creatures. In 2012, Dutch scientists found that mussels exposed to tiny plastics ate less of the algae that comprised their normal food – and grew less well – than mussels that hadn't been exposed. A 2016 study showed that oysters' reproduction rates nearly halved when they ingested plastic particles, mistaking them for phytoplankton. Another study that year showed that fish still in their larval state that were exposed to microplastics had reduced rates of hatching and development to maturity. The fish also ignored chemical signs that would ordinarily warn them about the presence of predators.

The microplastics weren't just remaining in small sea creatures. They were making their way up the food chain as larger creatures ate smaller ones, eventually ending up on the plates, and in the bodies, of human beings. In the years that followed, microplastics and nanoplastics – which are even smaller, measuring from 0.001 to 0.1 micrometres, making them invisible to the naked eye – were found in salt, honey, teabags, beer, tap water and bottled water.

They were found in the soil in which our food is grown and in the air we breathe. It also became apparent that plastic packaging used for food and drinks released microplastics into its contents. Despite this, consumer goods companies limited their response to eliminating microbeads in toothpaste and body scrubs. Regulators were looking to ban the microbeads anyway, and there were replacements readily available, such as walnut shells or apricot kernels. The microbeads became a priority because they were highly visible to consumers in gels, scrubs and toothpastes and could 'hit brand reputation', explains Gavin Warner, who worked on sustainability for Unilever for six years until 2020 and marketed its tea, detergents and other products for nearly two decades before that.

Microplastics complicated everything. Nobody could draw defined boundaries around the problems posed by plastics any longer – they were everywhere. Recycling and reuse, the two main solutions long offered up to fix plastic waste, suddenly had new problems. Studies showed that plastic recycling plants were releasing microplastics, and that reusable plastic food containers – particularly when microwaved but even when stored at room temperature or refrigerated – could release billions of microplastics and nanoplastics.

In time, microplastics were found in human blood, breast milk, placentas, lungs, testes and the brain. Researchers showed that hazardous chemicals used to make plastics fire-resistant leached from microplastics into human sweat.

In 2022, a prominent London law firm described microplastics as a 'ticking time bomb for litigation'. Not long after, the lawsuits began to roll in. A proposed class action accused baby bottle makers of failing to warn parents that the bottles, when heated, expose infants to microplastics. Several others targeted water bottle makers, arguing that their claims of 'natural' water were deceptive since the water contained synthetic microplastics.

For a long time, it was unclear what the implications for human health were, if any. Then, in 2024, a study published in the *New England Journal of Medicine* linked the presence of microplastics to an increased risk of humans having a stroke or heart attack. The researchers arrived at this conclusion after studying plaque (a build-up of cholesterol lodged in the wall of an artery) removed from 257 patients undergoing surgery for a condition in which fatty deposits block normal blood flow. They kept tabs on the patients for the next 34 months. After adjusting for factors like age, body mass index and existing health conditions, they concluded that patients with microplastics and nanoplastics in their plaque were *four and a half times more at risk* of a heart attack or stroke than those without.

Antonio Ceriello, one of the lead researchers, theorises that the presence of plastics in plaque could make it more unstable and prone to rupture, leading to a blood clot that can cause a heart attack or stroke. Another hypothesis offered up by Ceriello – who is head of the diabetes department at IRCCS MultiMedica, a research hospital in Milan – is that the plastic induces thrombosis.

'Ours is the first evidence that links the presence of plastic to a higher rate of cardiovascular events,' he says. 'In my opinion, this is just the start of long-term research.' Ceriello – who, at the age of 70, is a grandfather – worries about the health impacts that microplastics will have on younger generations. He's been exposed for a few decades, but his grandchildren have been exposed since they were in the womb.

'On the one hand plastics have improved my quality of life a lot,' he says. 'But what if your life is shorter or you get a stroke? It's a very high price, in my opinion, for a good quality of life.'

*

Five months after Christine Figgener found the turtle off the coast of Costa Rica, a non-profit called the Ellen MacArthur Foundation published a report predicting that by 2050 there'd be more plastic than fish in the sea.

The calculation was somewhat dubious. It simply extrapolated a current estimate that there were 150 million tonnes of plastic waste in the ocean to predict how this would progress over the next few decades. But it grabbed headlines around the world, adding to what was fast becoming the biggest backlash against plastic waste since the *Mobro*'s unhappy voyage in 1987.

EMF was founded in 2010 on the Isle of Wight, off the south coast of England, by the yachtswoman Ellen MacArthur. Simply put, its mission is to push businesses to shift from being extractive to regenerative. In its report, the foundation proposed creating a 'new economy' for plastic, one that stopped its leakage into the environment and decoupled it from fossil fuels. Companies, said EMF, could save vast sums of money by reusing plastic packaging that was currently thrown away, worth up to $120 billion each year.

'The New Plastics Economy' report was funded in part by Unilever. Under its CEO, Paul Polman – a straight-talking Dutchman who had once dabbled with the idea of becoming a priest – Unilever had been working towards an ambitious goal set in 2010 to double revenue while halving its environmental impact, including waste, by 2020. To make good on his pledge, Polman needed to know how much waste Unilever was producing and what it was made of. The gargantuan data collection exercise that ensued revealed that packaging was the biggest slice of Unilever's waste. Of this, plastic was the most problematic.

To cut plastic waste, the company had been redesigning bottles to be less dense and had launched more concentrated formulations that required smaller bottles. Both moves saved Unilever money, but neither made a big dent in its overall plastic waste.

273

'The New Plastics Economy' was published in early 2016. Soon after, Gavin Warner – who had recently moved into a sustainability role aimed at keeping Unilever's environmental initiatives on track – was scrolling through the report when he came across an infographic. It showed that 14 per cent of plastic produced globally was collected for recycling and just 2 per cent was turned into packaging of a similar quality. The remainder was either downcycled into lower-quality items or lost during the recycling process. 'I looked at that and thought, *Wow, plastic will be the issue of our time*,' says Warner, who grew up in Durban and, in his pre-Unilever days, tracked aircraft for the South African air force from a mobile radar station in the bush.

Recycling rates for Unilever's packaging were low in part because its sachets, deodorant containers and handwash pumps had never been designed to be recycled. Warner offered up a new target he thought the company should pursue: making all its plastic packaging reusable, recyclable or compostable by 2025. By now, Unilever was experimenting with the solvent-based technology it hoped would allow it to recycle sachets. With nine years to go, Warner was convinced that Unilever had enough time to crack its sachet problem – whether by redesigning sachets to make them recyclable via conventional processes or throwing its weight behind the new recycling technology.

Polman embraced the target, recognising the need for Unilever to be more ambitious on plastics. 'The issue crept up quicker on us than we all realised,' he tells me. 'Because of all the imagery at the time, the consumer made a jump to be ahead of the companies and said, "This is totally unacceptable."'

On 14th January, 2017, on the eve of the World Economic Forum in Davos, Polman announced Unilever's bold new 2025 pledge. 'We are inviting the entire consumer goods industry to follow the same path,' he wrote in an accompanying editorial in the London-based *Times*. 'It's time we closed the loop.'

*

In April 2017, Nils Simon, a 35-year-old political scientist in Berlin, called for a legally binding global treaty to end plastic pollution. The previous summer, Simon – who had spent his career working on international agreements to manage the risks posed by chemicals – was having a slow few weeks at work. 'I began thinking, *Why is there no regime on plastics? It's so odd*,' he says. 'It's evidently a transnational issue – it's produced in diverse places, everyone uses it.'

At the time, plastics were being discussed at an international level but only within marine pollution agreements. Non-profits had long campaigned about plastic waste, using imagery of trapped dolphins and sea turtles. While the impact of plastic on sea life was nothing to be scoffed at, Simon felt that this focus had come at the expense of other issues, such as greenhouse gas emissions and the human health harms from making and using plastics.

'The New Plastics Economy' report had shown that the production of plastics made up 6 per cent of global oil consumption – the equivalent of the global aviation sector. It said that if plastics production kept growing as expected, it would account for 20 per cent of all oil consumption and 15 per cent of the world's annual carbon budget by 2050. The budget was created by tallying up the carbon the world could emit if it wanted to remain below a 2°C increase in global warming.

The hazardous chemicals used to make plastics, such as bisphenol A (BPA) and certain phthalates, had also been overlooked by many campaigners. Studies showed the chemicals leaked into the environment or migrated directly into the food and drink they were meant to protect. Back in 2016, few governments were talking about these facets of the plastics problem. The consumer goods companies were the ultimate decision-makers when it came to what packaging showed up on supermarket shelves. Yet they were continuing to limit their response to lightweighting plastics or saying that they'd raise recycling rates.

In a 2017 report he wrote for the Berlin-based Heinrich Böll Foundation, Simon highlighted the global nature of plastic waste including microplastics. He described plastic as 'a transnational problem that gets bigger the smaller it becomes'. He called for a treaty that required sustainably managing plastics throughout their life cycle – production, design, use, recycling and disposal – rather than focusing only on waste.

Simon's proposal was met with a mix of enthusiasm and scepticism by the two dozen or so people from WWF and other non-profits who attended the small launch event organised by the Böll Foundation at its Berlin offices. 'Those people who did not know that much about plastics said, "You can't have a whole treaty for just one substance like plastic",' remembers Simon. 'And all the people who know about plastics were like, "What, you want to pull all plastics into just one treaty?"'

Later that year, the public's growing distaste for plastic waste was further turbocharged by *Blue Planet 2*. Anchored by 91-year-old David Attenborough, the BBC documentary series featured breathtaking footage of how sea creatures live and interact with each other underwater. Made over a four-year period, it was the UK's most-watched TV show in 2017. In China, so many people downloaded it that it temporarily slowed down the internet. Its final episode showed, in graphic detail, how plastic was damaging living creatures. The public, by now having spent hours immersed in this underwater world, was stricken by what it saw.

'What I said to the crew right from the get-go was, "Don't frame out the bad stuff. This needs to feel real and of its time",' James

Honeyborne, the series' executive director, says. 'You could create a timeless piece that revelled in the glory and wonder and magic of the oceans and totally ignored the current situation.'

Blue Planet 2 described how young dolphins off the east coast of the US had been unexpectedly dying for unexplained reasons. Attenborough explained how the dolphins were likely eating smaller creatures that had ingested microplastics laden with toxic chemicals. The documentary showed a one-year-old dolphin lying dead on a stretcher.

Attenborough also explained how coral reefs are rapidly bleaching. They could be gone by the end of the century, he warned, due to the burning of fossil fuels which releases carbon dioxide, making the ocean more acidic.

A particularly saddening segment showed a dead albatross chick, discovered with its tiny stomach pierced by a plastic toothpick. The bird had inadvertently been fed plastic by its parent, who had mistaken it for food, explains Lucy Quinn, the zoologist who found it. 'Think about parents coming back and feeding litter to their children,' she says. 'I suppose it's that feeling of we are wholly responsible for that moment – it's not the bird's fault that it picked it up.'

*

Buoyed by the rising momentum around plastic waste, by early 2018 the Ellen MacArthur Foundation had begun drawing up a plan to get hundreds of companies and governments to sign a pledge that would build on the one Unilever had made. 'We were inspired by those conversations with Unilever,' says Sander Defruyt, a Belgian engineer by training who worked on plastics for the foundation. 'If they could do it, why couldn't others?'

In May 2018, Defruyt invited executives from more than 40 companies, along with some non-profits and government officials, to convene at Schloss Krickenbeck, a 13th-century castle near Düsseldorf in Germany that had been turned into a hotel and events space. The 100 or so people who showed up – from Unilever, Pepsi, Coca-Cola, Mars, Danone and other organisations – broke into smaller groups over two days to discuss how to improve recycling, redesign packaging and collect better data on waste. At night, they played pool or table tennis and sipped cocktails from open bars dotted around the leafy property.

At the event, Defruyt proposed a 'global declaration on plastics', soliciting feedback on what kind of targets this should include and how these should be defined. He told a packed room that companies signing the voluntary declaration would need to agree to eliminate 'unnecessary' or 'problematic'[2] plastic packaging, increase recycled content and make all their plastic packaging reusable, recyclable or compostable by 2025. To fulfil this last bit, companies couldn't just make their packaging technically recyclable, reusable or compostable. This had to happen in practice and at scale.

Many companies were reluctant to sign the pledge. Ensuring a package was recycled in practice meant companies needed to get involved in expanding collection infrastructure, a responsibility that historically had been shouldered by municipal governments. But pressure was mounting on all the companies that had travelled to

---

[2] Some might argue that plastic packaging is inherently problematic, and indeed EMF's wide-ranging definition of problematic plastics could easily be interpreted to encompass all plastic. It describes problematic plastics as ones that aren't reusable, recyclable or compostable; that contain hazardous chemicals; that can be replaced with reusable options without diminishing performance; that have a high likelihood of being littered; and that hinder or disrupt the recyclability or compostability of other items. Companies interpreted this far more narrowly and mainly focused on phasing out 'black' plastics that can't be detected by optical sorters in recycling plants, PVC and expanded polystyrene.

Schloss Krickenbeck that day to show their products were in fact being recycled.

Only a few months earlier, an arrangement with China that for decades had allowed Western countries to burnish their plastics recycling rates had fallen apart. China had long been the world's largest buyer of scrap. By 2012, it was taking 58 per cent of all plastic waste exported by the US. Ships from China that had dropped off their cargo offered deep discounts to avoid returning empty. As always, it was money that dictated the arrangement; sorting recyclables was far cheaper in China than in the US and other developed countries.

But much of what the Western world sent east was hugely contaminated, with purportedly recyclable plastic waste often mixed in with household garbage and even medical waste. Single-stream recycling had exacerbated the contamination; the behaviour had become so bad that recyclers developed a new term for it: 'wishcycling'.

Workers in China lived alongside the scrap with their children, who played among the used plastic. On a 1992 visit to a plastics recycling facility in the Guangzhou countryside, Annie Leonard of Greenpeace described how plastic waste was sorted in a walled courtyard where workers also lived. 'A massive pile of discards – unrecyclable plastics, clothing, scraps, and other garbage – occupy the centre of the courtyard,' she wrote. 'The facility manager explains that there is no central dump in which this material can be disposed, so it is dumped in random locations in the countryside.'

Through the 2000s, the Chinese government's frustration was building. It had implemented tighter import controls and increased inspections, but contaminated and banned waste kept flowing in. Criticism began to mount domestically, particularly after a 2016 documentary, *Plastic China*, depicted the difficult lives of two families who sorted imported plastic waste.

Eventually, China's government had enough. In January 2018, it instituted a ban on 24 categories of waste. These included plastics and mixed paper. At first, the US, UK and other long-time exporters began rerouting their 'recycling' to other Asian countries. Those countries, including Malaysia, India and Thailand, instituted their own restrictions. By 2019, the US's plastic waste exports had dropped 69 per cent from 2014 levels.

In the months that followed, plastic and mixed paper waste piled up in collection centres across wealthy countries. Prices for used material crashed. It was a stark reminder that, just because something is *collected* for recycling, it doesn't mean it *is* recycled. As McDonald's and P&G discovered early on when they tried to recycle low-value, food-soiled Styrofoam containers and smelly disposable diapers – no buyer, no recycling.

In Philadelphia – whose single-stream programme Kate Krebs had once held up as a model of kerbside recycling – the city resorted to burning half its recyclables. While previously it had sold these for $67 a ton, it now had to *pay* $78 a ton for waste companies to take its recyclables off its hands. 'We had no alternative but to get into this type of arrangement,' Carlton Williams, the city's Department of Streets Commissioner, told me back in 2019. Williams blamed the situation on how contaminated Philadelphia's recycling had become, with residents putting everything into one bin. Other municipalities, faced with having to pay to recycle material they once sold, cancelled recycling programmes altogether.

Just as the *Mobro* had once shown America that its take–use–throw culture was getting out of control, China's ban reminded the developed world that what many had long believed was a workable recycling system was in fact a house of cards. The economics of recycling most plastics didn't line up – they never had – and shipping mixed plastic waste off to China in the name of 'recycling', when in fact a big chunk was dumped or burned, could no longer be passed off as a feasible way of doing things.

In the US, the head of a plastics recycling trade body warned that China's ban could lead to 'a tsunami' of 'very stringent' regulations. 'The brands are going to have to show – in my opinion – sustained commitment to incorporating recycled material in their marketplace, or I believe that they're going to be forced to do it,' Steve Alexander, president of the Association of Plastics Recyclers, told members.

The brands decided the time was ripe to show commitment. A year later, in October 2018, Ellen MacArthur announced that 250 organisations had agreed to make all their plastic packaging reusable, recyclable or compostable by 2025. The signatories, which included companies like Pepsi, Coca-Cola, Danone, H&M, L'Oréal and Mars – in addition, of course, to Unilever – had also agreed to cut unnecessary plastic packaging, to use more recycled plastic, to phase out problematic plastics and to explore reusable packaging. Together, the companies accounted for 20 per cent of plastic packaging sold globally.

At a time when the world was desperate to see evidence of action on plastic waste, EMF's plastics pact was hailed as a game-changer. The United Nations Environment Program described it as the 'most ambitious set of targets we have seen yet in the fight to beat plastics pollution'.

*

If you've read this far, you already know – or at least suspect – what happens next.

Companies did not make all their plastic packaging reusable, recyclable or compostable by 2025. Most didn't even come close.

Unilever was among the first to announce it was stepping away from the 2025 target.[3] Paul Polman was long gone and in his place was a new CEO, Hein Schumacher. He dismissed the previous plastics targets as unrealistic, vowing he would reshape the company to be more performance-focused.

Schumacher's announcement sparked an uproar. At the company's annual shareholder meeting at the Hilton hotel in south London in May 2024, Greenpeace activists staged a protest, letting off confetti guns. 'You and your board are plastic polluters,' shouted an activist before security marched her out. 'You are responsible for the mess that we're in today.'

While Unilever wasn't the first big company to step away from its commitments – Nestlé, Pepsi and Danone had done so months earlier – many environmentalists and employees felt betrayed and anxious. Despite its sachets, they had perceived Unilever as the most environmentally conscious of the major consumer goods companies. If Unilever was saying it couldn't do it, what hope did the rest of the industry have?

'If you water down targets, that's you saying "This is what I can do" versus "This is what needs to be done",' says Kavita Prakash Mani, who served on Unilever's sustainability advisory council until the company disbanded the body ahead of announcing that it wouldn't meet its plastics targets. 'You can say it's pragmatic. It's disappointing. We all feel a bit sad.'

For anyone paying attention, the writing had been on the wall since at least 2020, when the excitement of having the world's largest brands sign up to the plastics pledge had died down. The realities of needing to deliver had sunk in. In EMF's annual progress report on

---

[3] Unilever shifted its goal, saying all its rigid plastic packaging would be reusable, recyclable or compostable by 2030. It said its flexible plastics would hit this target by 2035.

the pledge that year, Sander Defruyt had warned that brands needed to do much more to reduce flexible plastic and move to reusable packaging if they were going to meet their targets.

The Covid pandemic that hit in early 2020 upended supply chains, whipsawed commodity prices and propelled huge swings in consumer demand. It also briefly paused recycling efforts. All this raised enormous challenges for companies. But Covid wasn't why the consumer goods makers were struggling to hit their targets. Rather, the fundamental problems companies had faced since the very inception of plastics recycling remained unsolved.

Virgin plastic was generally still cheaper and better than recycled plastic. Sorting dozens of different types of plastics – combined with each other and with other materials – was very expensive. The only plastics recycled in large volumes were bottles and jugs made from PET and HDPE – the same two easily recognisable plastic containers the industry had begun recycling in the 1980s. There had been little progress on recycling everything else. Buyers for most used plastics just didn't exist. As a result, the infrastructure to collect and sort plastics was sorely lacking.

Conventional 'mechanical' recycling involves cleaning, shredding, melting and re-extruding plastic. Plastic degrades during the process, meaning it's usually only recycled a couple of times. Pigments can't be eliminated and sorting by colour is expensive, so coloured plastic gets downcycled into grey pipes or building material. Shape matters, too: trays and clamshells have different properties to bottles and jars made from the same plastic. This means they can't always be recycled together and are often landfilled or burned.

Mechanical recycling presents added problems when it comes to recycling packaging used for food, particularly flexible packaging like films and bags. These are more sensitive than rigid plastics to contaminants like dirt or oil and are more dependent on additives – which can degrade during the recycling process – to give them properties like flexibility and strength. While all plastics are hard to clean and recycle into a high-quality plastic that's consistently safe enough to be used for food and drinks, flexible plastics are particularly hard. Imagine trying to wash a greasy curry out of a plastic bag – enough said.

Many of the consumer goods companies that had signed the EMF pledge had been trying to deal with these limitations by investing in chemical recycling. The term encompasses an array of technologies that use chemicals or heat to break down plastic into its basic chemical building blocks so it can be turned into clean, new plastic that's high-quality enough to make new food packaging again and again. Progress had been slow. The various technologies all sounded amazing on paper, but for decades had been unable to scale, held back by high costs and questionable environmental credentials. I'll say more about chemical recycling in the FAQs.

Back in 2020, although some big companies had signed up to the EMF commitment, many hadn't. Defruyt says EMF had wrongly assumed a big ramp-up in participation and that the signatory group's increased use of recycled content would spur investment to expand collection and sorting infrastructure.

Instead, the consumer goods companies that had signed the EMF pledge found themselves at a disadvantage. They were competing against rivals who weren't spending the money to switch away from plastic or paying more to use recycled plastic. Unpredictable legislation presented a further disincentive to try new technologies or redesign packaging since recyclable alternatives could suddenly be banned.

Compostable and biodegradable consumer packaging was proving to be a largely nonsensical claim in real life. Most people don't compost dozens of bags and cups in their gardens, and most industrial composting facilities don't accept compostable packaging (I'll explain why in the FAQs). Reusable packaging sounded great on paper, but without the infrastructure for companies to take reusables back, and the regulation to make them as cheap and convenient as disposables, reusables had remained a pipe dream.

A lack of data and conflicting definitions posed other problems. Only 68 per cent of countries had a national waste agency at all, and only 39 per cent publicly reported waste data. There were seven different definitions around the world of what a single-use plastic bag was, depending on micron thickness. These variations meant that getting a handle on plastic waste and reporting progress was difficult, yet another reason why consumer goods companies – for whom sustainability has always been a marketing tool – were loath to invest.

A 2020 analysis showed that current commitments by governments and industry would only reduce the annual volume of plastic flowing into the ocean by about 7 per cent by 2040. It was nowhere near enough.

Since Nils Simon's report, the idea of a global plastics treaty along the lines of the 1987 Montreal Protocol – which banned chlorofluorocarbons – or the 2015 Paris agreement on climate change had caught on. Law professors, policy experts and environmental non-profits were lobbying for a treaty and even the American Chemistry Council – which had absorbed the American Plastics Council – said it was in favour of a treaty, so long as this was focused on cutting plastic *waste*.

Then, in January 2022, Coca-Cola, Unilever, Pepsi, Walmart and a handful of other consumer goods and retail companies publicly announced that they supported the idea of a plastics treaty that would lay down global rules all businesses had to follow. Unlike the American Chemistry Council, the consumer goods companies were calling for a treaty that addressed all aspects of plastics – from its production and use to its disposal and reuse. 'For companies and investors this creates a level playing field,' said the companies in a statement announcing their support.

Like drug addicts who were finally ready to acknowledge that their repeated efforts to voluntarily stop using would never work, the companies that had long been so hooked on plastic were now asking to be taken to rehab. As long as all their peers were sent too, they'd accept it, they seemed to be saying.

Finally, negotiations for a global plastics treaty kicked off in Uruguay in November 2022 with 2,300 people from 160 countries and organisations attending. The consumer goods companies attended under a group they called the Business Coalition for a Global Plastics Treaty. They said a treaty should require businesses and governments to make changes that would reduce plastic production, improve recycling and composting, and tackle the release of microplastics.

Without global rules in these areas and others, the companies had calculated that they were staring down the barrel of 400 to 600 new local or national laws worldwide each year that took aim at the environmental issues caused by plastics. The bills were often focused on single products, like bags or straws, and were restricted to a local jurisdiction. They had little chance of making a big dent in the plastic waste that was threatening the companies' social licence to operate.

Darren Woods, the CEO of ExxonMobil – the world's largest maker of plastics – reassured his investors that such siloed restrictions on bags and straws posed little threat to the oil company's $36 billion in annual sales. Substitutes for plastics, like paper or seaweed, were expensive and their supply was constrained, he said in May 2024. An unwarranted level of attention had been given to a handful of bans focused on plastic straws, cutlery and takeaway containers, said Woods. 'Current bans are location-specific and don't represent a significant impact on our business,' he told investors. 'It wouldn't even if bans were enacted more broadly.' He confidently described a significant reduction in plastic demand as 'an unrealistic future scenario'.

The Exxon CEO had, unintentionally, offered up the clearest argument yet as to why a total sea-change in policy and approach to the world's increasingly multifaceted plastics problems was needed. Fragmented bans weren't working, global plastic consumption and waste were both forecast to nearly triple by 2060, and it was clear that recycling systems were ill-equipped, overwhelmed and unable to deal with even the existing flood of plastics coming through their doors.

*

In March 2024, yet another report on plastics was published. This one wasn't about waste. Instead, it explained that plastics can contain thousands of potentially harmful chemicals and that these can leach into our food, our homes and the environment.

It was six weeks before the penultimate negotiating session for the plastics treaty and the stakes were high. The report – funded by the Norwegian Research Council, a unit of Norway's government – called on governments engaging in the treaty negotiations to push companies to use far fewer and safer chemicals, to tighten regulation and to require much more transparency about the chemicals used to make plastics.

Plastics, as I mentioned in Chapter 1, are made up of chemicals: long chains of monomers like ethylene, propylene, styrene and vinyl chloride. Chemicals are also used to start or speed up polymerisation, the process of linking monomers. Chemical additives give plastics flexibility, resilience, colour and other properties. They can make up between 0.05 per cent and 70 per cent of a plastic by weight. As the British Plastics Federation explains on its website, 'without additives, plastics would not work'.

But many of the chemicals found in plastics aren't intentionally added. The consumer goods companies – and even the plastic packaging makers who sell to them – don't know exactly what's in plastic packaging.

When I first heard this, I found it so baffling that I needed it explained to me in the most basic terms possible. I phoned Jane Muncke, an environmental toxicologist who is one of the 2024 report's authors, and asked her how such a thing was possible. 'Think about when you buy a bunch of spinach,' Muncke explained. 'Often it's gritty and comes with sand or dirt.' Similarly, she said, when companies buy monomers to make their plastics, these are often contaminated with other chemicals that can find their way into the final packaging.

Non-intentionally added substances can also come from the polymerisation process, which produces by-products, or from the degradation of additives. Studies have found that they can also sneak into plastics during the recycling process. These non-intentionally added substances can account for up to half the chemicals found in plastics, but are hard to test for. Even when detected, they can be very hard to measure. Muncke explained that scientists need pure chemicals to calibrate their instruments. Having this baseline measurement allows them to determine the concentrations of chemicals in packaging or food. But if scientists can't buy pure chemicals on the market, they can't calibrate their instruments, which means they can't measure the chemicals.

Workers are exposed to chemicals while extracting fossil fuels and during the production of plastics. The rest of us are exposed while using plastics. Chemicals can also escape after use – from plastics that are dumped or burned, leaching into the oceans and the soil. They accumulate in microplastics, as I outlined earlier, which can then be ingested by sea creatures and, eventually, by humans.

The researchers of the 2024 Norway-funded report identified 16,000 chemicals as present or potentially used in plastics. Of these, more than 4,200 chemicals are 'concerning' because they persist in air, soil or water for long periods, accumulate in human or animal bodies, are toxic, or spread in fresh or drinking water. Only 6 per cent of the 16,000 chemicals are regulated globally.

The researchers flagged major data gaps, noting that basic safety information about the potential hazards posed by thousands of chemicals is missing. The impact of mixtures of chemicals used in plastics on human health also isn't considered. Regulation is aimed at individual chemicals, yet we're all exposed to multiple chemicals at once every day.

'We simply do not have enough information on the plastics that we're using on a daily basis,' says Martin Wagner, a German biologist who researches how plastics impact our health and was the report's lead author. 'It's been either protected as confidential business information or the chemicals are unknown.' Wagner says the many thousands of chemicals found in plastics can't possibly be adequately regulated because no regulatory body has enough resources to assess them all. He, Muncke and others I spoke with say that even where rules do exist, enforcement is lax and post-market testing of products on shelves is virtually non-existent.

For decades now, hundreds of studies have shown that dangerous chemicals are leaching from plastic into food and drinks. Sometimes this happens at levels well above safety standards, which are themselves controversial. One of the most notorious chemicals associated with plastics is bisphenol A. Widely found in food and drink can liners and in reusable water bottles, BPA has been linked to an increased risk of cancers, impaired immune function, early puberty, obesity, diabetes, anxiety, depression, impaired brain development and hyperactivity.

Another hazardous chemical is vinyl chloride, which is used to make polyvinyl chloride, a plastic used for pill packaging, water pipes and electrical cable insulation among other things. As early as the 1970s, vinyl chloride was linked to a rare liver cancer and to abnormally high miscarriage rates. Today, the National Cancer Institute – the US government's main body for cancer research – says exposure to vinyl chloride is linked to brain and lung cancers, lymphoma and leukaemia.

Then there are phthalates, a class of chemicals commonly used to soften plastics. Phthalates are found in toys, detergents, scented candles, blood bags, inks, adhesives and lots of other products. Studies have shown that certain phthalates raise the risk of infertility, early puberty, miscarriage and premature birth. They've also been linked to neurodevelopmental problems.

More recently, PFAs – originally found to be widespread in greased paper used to make food packaging – have also been found in plastics. The PFAs – as I mentioned in Chapter 3 – have been dubbed 'forever' chemicals for their ability to hang around in nature and in our bodies. They have been linked to immune system suppression, higher cancer risk and liver problems.

Regulators have set safe migration limits for some of the chemicals that are intentionally included in plastics, meaning these can leach into food and drink but only under certain quantities. But many researchers criticise the limits as too high and based on outdated evidence. In Europe in 2023, the food safety regulator recommended lowering the tolerable daily intake for BPA by *20,000 times* from the previous level set in 2015.

In fact, many researchers say there are *no* safe levels of such chemicals, which they call endocrine disruptors. These chemicals, even in tiny amounts, have been shown to mimic or otherwise disrupt the hormones essential for reproduction, growth and a range of other crucial functions. A large number of scientists believe that endocrine disrupting chemicals are playing a part in declining fertility rates around the world. 'The power of these chemicals to impact fertility is mind-boggling,' says Patricia Hunt, a professor at the School of Molecular Biosciences at Washington State University who specialises in studying chromosomally abnormal human eggs. 'We have all sorts of evidence that indicates "Whoa, we're in serious trouble here."'

Concerning as they are, BPA, PFAs and phthalates have been widely studied. The plastics treaty negotiations opened the door for researchers to highlight that there are thousands of other chemicals that are less well studied and that current regulation to protect human health from chemicals is ineffective and insufficient. It was a message they had delivered for decades. But chemicals are invisible to regular people and their impacts are hard to pin down. They have rarely sparked the same kind of outcry as plastic waste.

As the treaty negotiations unfolded, a group of academics and researchers named the Scientists' Coalition for an Effective Plastics Treaty called for a ban on chemicals widely established as hazardous. Beyond the individual chemical, they also wanted the chemical's entire class banned. It sounds draconian until you realise that companies' response to concerns about bisphenol A has often been simply to move to bisphenol S, which belatedly has been found to cause similar harms. In other words, a BPA-free bottle may still contain bisphenols. Scientists call such stand-ins 'regrettable substitutes' and say that banning chemicals by individual type is pretty futile, akin to playing whack-a-mole.

The scientists' coalition also wanted a treaty to draw up a global list of all the chemicals used in plastics and to divide these into positive and negative chemicals, with the harmful ones in time phased out. When it comes to the thousands of non-intentionally added chemicals, Martin Wagner – who is a member of the coalition – says researchers know little about their structure or the risks they pose. Companies, he says, need to redesign packaging so that it doesn't contain as many chemicals and must find ways to change manufacturing and recycling processes to stop unintentionally added ones from being introduced.

'If, through a plastics treaty, governments could agree to take the worst chemicals off the market, that would be significant,' says Wagner. 'But given there could be more than 16,000 chemicals in plastics, that's really just a drip on the stone as we say in German.'

*

'I ask myself a million times the question of how we got here,' says Pablo Costa. 'Nobody saw this coming. This wasn't planned, it wasn't intentional.'

It's a blustery February day up in North Yorkshire, where Costa – who is Unilever's global head of packaging – is on the remote farm he calls home, a few miles from the Yorkshire Dales National Park. He's perched on a chair near a window and on the wall behind him hangs a sunny, seaside photo taken somewhere in Spain, a reminder of better times that have been or a symbol, perhaps, of better ones yet to come.

I've asked for Costa's opinion on how we arrived at a point where our world is awash with single-use plastic and why none of the many efforts that companies have undertaken to rein plastic in over the past four decades since the *Mobro* have worked. 'The biggest mistake we made on plastic is we got the business case wrong, because it's so cheap,' says Costa. 'Anything we try to do is more expensive.'

Plastic, Costa tells me – warming to his theme – is too good at its job. It's so light, so moisture-proof, so tough, such a great barrier to oxygen, light and grease, and it does all this at such an affordable price that no other material comes close. Companies that at first tentatively tried using plastic loved it so much that they embraced it with open arms. 'This is how we fell into this trap,' he says conclusively. 'We started abusing plastic.'

The trap Costa says big brands have stumbled into is easy to see when you look backwards.

McDonald's still sells billions of plastic-lined paper cups a year, only a tiny slice of which are recycled. Its reliance on disposability has left it struggling to convince customers there for the fast food to slow down enough to scrape leftover fries out of their paper containers so these can be recycled. Across the rest of the industry, foam fast-food containers are still widely used. Lawmakers continue to push for polystyrene bans while the plastics industry continues to warn of price rises and claim the containers can be recycled. If you ignored the dates on present-day news articles on the issue, you'd easily believe they were written in the late 1980s.

Procter & Gamble still hasn't found a way to make the economics of recycling diapers work. Between 2011 and 2018, a large global study of 82 countries found disposable diapers were among the top 25 most littered items on the seafloor and among the top 40 most littered items on land. Every minute, more than 300,000 disposable diapers are landfilled, burned or littered around the world. The disposable diaper industry uses more than 248 million barrels of crude oil each year.

Hefty never relaunched the degradable trash bags that got it into hot water, but it has run into trouble for another eco-friendly claim. In the summer of 2024, it settled a lawsuit from the Minnesota attorney general who alleged that the brand misled consumers by marketing its recycling bags as 'recyclable', when in fact putting recyclables into the bags ensures they and everything in them go to landfills. Hefty settled a class action lawsuit that made similar allegations in 2023. It's currently facing a lawsuit from Connecticut's attorney general over the bags.[4]

Meanwhile, the North American market for trash bags is projected to grow at a healthy 7.7 per cent a year in the five years to 2029 as people and companies keep buying more stuff and throwing it away.

Degradability continues to be a widely abused claim, despite several updates to the FTC's 'Green Guides' and the implementation of guidelines in other countries. Hundreds of companies are marketing their wipes, diapers, straws, sponges, cutlery and dog poop bags as biodegradable or compostable without telling consumers that most compost facilities don't take these, and that in landfills they'll either remain mummified for decades or break down and release methane.

The chasing arrows symbol is still widely used on plastic products for which there are no recycling markets. The 'Green Guides' continue to advise marketers that, as long as their products can be *collected* for recycling, they can label them recyclable regardless of whether they are actually turned into new products. Of all the plastics that consumers in the US throw away, just 5 to 6 per cent are recycled, according to a 2022 estimate. As I mentioned in Chapter 7, 92 per cent of Americans say they don't know what the chasing arrows symbol and the numbers it contains mean.

Unilever has not found a way to collect the trillions of sachets it sells each year for recycling. Nor has it found an alternative to replace these. Paul Polman says he regrets not pushing the company harder on sachets during his decade as CEO. 'We tried to find solutions to the problem whilst our mindset should have

---

[4] In an emailed statement, Reynolds Consumer Products – Hefty's current owner – said it doesn't comment on pending litigation, but that its product labelling 'has never claimed the bags themselves are recyclable'.

been that we should not create the problem in the first place,' he told me.

Coke continues to hold up recycling as its main strategy to cut waste and Kate Krebs is now head of external affairs for Closed Loop Partners, which works closely with Coke and other drinks makers to improve recycling. In 2019, together with its soft drink counterparts at the American Beverage Association and Closed Loop Partners, Coke launched an advertising campaign called 'Every Bottle Back'. It bore strong similarities to the 'Give it Back' campaign from a decade earlier. It was deemed misleading by the US's National Advertising Review Board, which ruled that statements made in the new campaign overstated the extent to which recycled materials were currently used in products.

In December 2024 Coke, once again, watered down its recycling targets, saying it would use 35 to 40 per cent recycled material in bottles and cans by 2035. It had previously pledged it would use at least 50 per cent recycled material by 2030. Fewer than 12 per cent of the PET bottles sold in the US today are recycled into new drinks bottles.

Coke also never hit its 2020 plant bottle target. When I enquired about it in 2024, the company's senior director of sustainability communications seemed confused. 'There was never such a goal,' she wrote. 'Let me know if you can share where you got this from.' I sent her two old press releases from Coke referencing it, along with a 2011 *New York Times* article in which the company said it would make all its plastic bottles from plant-based plastic by 2020. She never replied.

Latest available data shows that reusable packaging makes up just 14 per cent of Coke's sales. In December 2024, it quietly dropped a target it had previously set for refillables. It didn't set a new one.

In June 2024, the city of Baltimore in Maryland filed a lawsuit against Coke and Pepsi, saying the companies had significant roles in creating a plastic pollution crisis. The mayor's office pointed to plastic's abysmal recycling rates, but also to the microplastics found in human bodies and the chemicals that can leach out of littered plastics. Four months later, Los Angeles County filed its own lawsuit against the two drinks makers, saying they had misled consumers into thinking recycling would fix the problems caused by their disposable plastic bottles.

Also in 2024, California's attorney general sued ExxonMobil for a 'decades-long campaign of deception that caused and exacerbated the global plastics pollution crisis' and for misleading consumers into believing recycling could solve the plastic waste crisis when it 'clearly knew this wasn't possible'. Exxon in response sued the AG and several environmental groups, saying they were engaging in a smear campaign.

It all sounds pretty familiar, doesn't it? When I look back on the years since the *Mobro*'s trash was burned, it's clear to me – and hopefully now to you as well – that some very fundamental things need to change in order for companies to get off the hamster wheel they're on, making and breaking the same promises over and over again while operating using the same flawed assumptions within the same self-serving linear system they helped create.

Doing so is trickier than it's ever been. In December 2024, global talks for a plastics treaty collapsed. Countries couldn't agree on key issues like whether to cap plastics production and phase out chemicals of concern. In the meantime, our plastic problem continues to spiral well beyond where it began – omnipresent microplastics and chemicals have made the picture far more complex, and any efforts to tackle these issues must also ensure that alternatives don't cause higher carbon emissions. It's enough to make anyone's head spin. Addressing one problem without unintentionally making another worse will require a careful balancing act.

CHAPTER 13

THE FUTURE OF PLASTICS IS NOT IN THE TRASH CAN

'We need legislation'

The first thing that struck me as I walked into the Carrefour store on rue de Grenelle was that all the cucumbers were naked.

Nestled in wooden crates, they reflected the store's ceiling lights and looked as inviting as any cucumbers could on a chilly autumn evening. It was November 2023 and I had taken the train from London to Paris to meet the French supermarket giant's sustainability head, Bertrand Swiderski.

Starting a few months earlier, France had outlawed plastic packaging for a wide range of fruits and vegetables. I was curious about Carrefour's learnings so far. If a mainstream supermarket could eliminate plastic for a range of fruit and vegetables without causing any big knock-on effects, surely this could inform the way grocery chains in other countries package their produce.

Swiderski – a lanky, fast-talking Frenchman who has been Carrefour's sustainability head for more than a decade – told me that the chain made minor adjustments to its supply chain, reducing the number of days the cucumbers spent in warehouses and transit so they arrived on shelves more quickly. The cucumbers looked in good condition despite eschewing plastic and Swiderski said Carrefour wasn't seeing more waste. 'I think it was a lobby on plastic trying to keep plastic on cucumbers because it's huge,' he said conspiratorially, waving expansively at the pile of unwrapped cucumbers.

It remains unclear to me whether we should push for cucumbers to be sold naked everywhere. Plastic advocates often point to the cucumber as the best example of a product that benefits from plastic. It takes 93 plastic cucumber wraps to have the equivalent environmental impact of one cucumber thrown away. In other words, the wrap is much less environmentally wasteful than a spoiled cucumber. Studies conflict on whether plastic-wrapped cucumbers stay fresh for longer than unwrapped ones.[1] However, one big takeaway from Carrefour's changes is that supply chains aren't set in stone. They can be rerouted and shortened, allowing us to change packaging types or use less packaging.

To minimise food waste, companies could source more locally, sell what's in season, more tightly manage their inventory, and stock fewer products that sell more quickly. Often supply chains are only as lengthy and convoluted as they are because plastic packaging has *enabled* them to be this way.

---

[1] A 2011 study in the *Journal of Food Science & Technology* said that plastic-wrapped cucumbers stay fresh for 60 per cent longer than their unwrapped brethren. However, WRAP, a UK waste reduction non-profit, says its testing shows plastic wrap doesn't increase the shelf life of cucumbers; this is true whether stored in fridges or at ambient temperatures. However, it also told me that suppliers sent in feedback saying plastic wrap does reduce bruising through the supply chain.

Back at the store, Swiderski and I continued on our loop of the fresh food section. Carrefour has continued to package much of its produce, but has switched to paper. It was using stiff cardboard containers to hold half a dozen apples, pears and bunches of cherry tomatoes. Swiderski explained that the packaging doesn't extend shelf life,[2] but allows Carrefour to sell the products in bulk at a fixed price.

Studies show that such bulk packaging, although more convenient and usually cheaper per unit, can increase food waste since it forces people to buy more than they need. In the UK, where much of what's sold in supermarkets is wrapped, 60 per cent of food waste happens at home.[3] Fresh vegetables and salads are the most wasted category. WRAP says that selling items like bananas, potatoes and apples unwrapped is the single most important factor in reducing wastage of these foods, since people can buy the amount they need.

Many products at the Carrefour store were still wrapped in plastic. These included bags of large carrots. I asked Swiderski why Carrefour had eliminated plastic for its cucumbers but not its carrots. 'We do not remove plastic because we don't know how to do it,' he said, shaking his head sadly. Surprised, I pressed him further. 'We didn't find a solution with carrots,' he insisted firmly. 'Look,' he continued, grabbing a bag of carrots and holding it up. 'Look. Because of the humidity we need to keep inside, we didn't find any solution.'

We rounded the corner of the vegetable display and both stopped. There in front of us sat a large pile of unwrapped carrots. 'What about those carrots?' I asked. He paused. 'We've got both. Always we've got both. We've got a bulk solution and we've got this one also.'

---

[2] Shelf life just means how long a product remains fresh enough to eat.
[3] I outline a few suggestions for how we can cut food waste at home in the FAQs.

'Why can't you do the same thing you've done with bananas and use a paper band to wrap them?' I asked. 'Why not', he said. 'Why not.'

'If it were a humidity issue, you wouldn't be selling those,' I continued, pointing at the unwrapped carrots. 'That's true,' he said, looking only slightly chastened. 'No, no, it could be done.'

Having learned the hard way – as the mother of two toddlers – that it's best to pick my battles, I let this one slide. Later, I deduced that the carrot–cucumber discrepancy likely boiled down to Carrefour wanting to distinguish between its organic carrots and ones that use insecticides. Swiderski explained that Carrefour still wraps organic avocados in net bags but leaves their cheaper counterparts unwrapped just so it can distinguish between them at the check-out. Retailers also choose to wrap products so they can tell people where things are from and offer information about storage and cooking methods, some of which is mandated by regulation.

Both these issues seem pretty solvable as technology improves. Compostable stickers bearing QR codes, laser embossed symbols, and cameras that use artificial intelligence could help automated check-out machines and store staff distinguish between various types of avocados. The QR codes – along with old-school signs on shelves – could offer people information too. Smart check-out systems that auto-weigh and auto-recognise produce could also greatly lessen the inconvenience of buying produce loose. As with supply chains, labelling requirements have evolved to cater to disposable packaging, but things could be done very differently.

Less surprisingly, Carrefour was also selling all its berries wrapped in plastic. Berries are exempt from the French law and scrapping plastic for them would be 'impossible', Swiderski said.

After saying goodbye to Swiderski, I walked back towards the Seine. On impulse, I stopped at Alimentation BAC, a small fruit and vegetable store, all of whose berries were displayed plastic-free in open cardboard cartons. Swallowing my awkwardness, I attempted to ask the shop owner why he didn't feel the need for plastic packaging. An English-speaking customer eventually took pity and translated my question. The owner guffawed. He told us that the store gets fresh deliveries each morning. Everything we saw would be sold out by the end of the day, he added.

*

You'd think cutting back on unnecessary packaging and reducing complexity would be in brands' best interests but, for many decades now, a package has been about far more than simply protecting its contents.

Ever since the National Biscuit Company launched what many say is the first branded consumer package in 1899 – putting its Uneeda soda crackers in an envelope of waxed paper slipped inside a cardboard box – brands have seen packaging as a marketing tool. By the 1930s, honey was being sold in 36 different kinds of containers in New York City alone. Nineteen of those containers accounted for only 4 per cent of the honey sold. On the other hand, more than 50 per cent of the honey sold was in containers of just three types and sizes.

The 'demand' for such unnecessary convenience and choice didn't exist; most people were happy to buy their honey in just a few types of containers. 'Packaging has been carried out in the disorderly, unplanned fashion so characteristic of the hurrying, heedless America,' complained an editor for the US Department of Agriculture in 1938.

The 1930s, as you may recall from Chapter 1, were when retailers first embraced plastic in a big way. There were undeniably some environmental benefits. It wasn't only that plastic wrap kept meat or broccoli fresh for longer. Pre-trimming and pre-packaging cauliflower and lettuce meant more could fit on trucks and ships, while leftover stems and leaves could be used for animal feed, composted or repurposed by manufacturers to make soup or coleslaw. Using plastic also allowed store-owners to slash costs since they didn't need staff to weigh and price produce.

But a major reason supermarkets took to plastic packaging with such enthusiasm was that plastic encouraged people to buy more.

Soon after the Second World War, a plastics industry executive wrote of the difference between 'utility' packages – cardboard boxes, glass bottles and tin cans designed to contain or protect a product – and plastic wrapping films, which offered utility but, moreover, visibility. 'We are interested in the eye appeal, the attractiveness, the stimulus to impulse buying which transparent wrapping film can impart to many lines of consumer goods,' wrote Francis E. Simmons of the American Viscose Corporation in a 1949 article entitled 'Packaging Aids to Marketing'. 'The use of transparent wrapping films in packaging is serving to bring about something of a minor revolution in the experience of marketing,' he wrote. 'This revolution is not one affecting the nature or form of the product but it is one which materially affects the presentation of the product to the consuming public.'

By the 1960s, it was clear that packaging was essential for turning commodities into brands. 'How do you stop the industry from hiding behind the consumer, which it does by inducing him to buy packages against his instincts and then declaring that the proliferation of packaging waste is only a response to "consumer demand", the very demand which industry itself has created?' asked one infuriated presenter at the US's first national conference on packaging waste in September 1969.

Companies have flooded the world with plastic with little thought given to what happens once it's used. They've sold us trillions of single-use sachets, pouches, films, diapers, bags, bottles, tubs, cups and other products in a baffling array of colours and combinations, sometimes incorporating chemicals that can harm human health. Virtually none of the products were originally designed to be recycled or reused. All were designed to promote greater consumption.

Unlike glass, aluminium and paper, plastic was invented in a lab and its various forms are seemingly endless. Plastics have been like an all-you-can-eat buffet with no stomach cramping or diarrhoea to deal with at the other end. The food is cheap and delicious. You can eat as much as you like, consequence-free. Who wouldn't take advantage of that?

The affordability of single-use plastic is an illusion, though. People like you and me have effectively subsidised plastic by shouldering billions of dollars of disposal, environmental and social costs that should rightfully be borne by the companies that have chosen to use it. It's time to end this subsidy and to ensure that the brands who choose to use plastic – and, indeed, all packaging – bear the *full* cost of doing so.

The 'polluter pays' principle, at its bare minimum, requires that companies cover the cost of ensuring their packaging is recycled or safely disposed of at the end of its life. In Germany, where the polluter pays principle – also called extended producer responsibility – first took hold in the early 1990s, the recycling rate for plastic packaging stood at 67.5 per cent in 2022,[4] which is well above most other countries. Germany has recycled more than 60 per cent of all its municipal waste each year since 2005.

A more effective version of the polluter pays principle levies fees on companies in a way that pushes up the price of more

---

[4] One reason why Germany hasn't made more progress is that its polluter pays model hasn't adopted tiered fees that incentivise a shift away from non-recyclable packaging. In the absence of waste-reduction targets, Germany has focused mainly on recycling.

environmentally burdensome packaging, incentivising brand-owners to make changes so that in time the percentage of waste flowing to landfills or incinerators reduces sharply. The idea is starting to catch on with more governments.

Financial penalties are among the strongest ways to change corporate and consumer behaviour. An example is the UK's 10-pence fee on single-use plastic bags, which acted as a circuit breaker for a deep-seated consumer habit, within a few years propelling a 98 per cent decline in single-use bags purchased.[5]

Laws that push companies to simplify, taking us back to a time when we had far fewer containers in fewer designs and fewer sizes that used fewer chemicals, would make it easier to safely recycle and reuse these. To be clear, I'm not advocating going back to the 1930s, nor am I suggesting that we eliminate all plastic packaging. There are clearly benefits to using plastic to package products like meat, a highly carbon-intensive product that spoils quickly if left unwrapped. But 100 years of 'disorderly, unplanned' packaging and product choices have got us to a point where it's clear we need a reset.

Getting the polluter to pay creates supply, but to spur demand it should be coupled with laws requiring minimum recycled content.

---

[5] While many people in the UK have developed a habit of bringing their own reusable bags when shopping, an unintended consequence of the single-use plastic bag charge has been that sales of thicker, multi-use plastic bags have increased. The number of bags sold isn't anywhere near that of the thinner plastic bags retailers previously handed out, but by some estimates overall plastic use has increased. Litter was the major problem caused by thin bags and the thicker bags aren't widely littered, but increasing plastic waste makes this far from an unadulterated win. The UK's experience shows that financial penalties do work to shift behaviour, but must be executed thoughtfully. Bags – paper or plastic – should be priced high enough that shoppers only buy them if absolutely needed. The price on thicker plastic bags could include a hefty refundable deposit to incentivise people to eventually return them to stores for recycling.

This would create large, consistent markets for a number of used plastics and incentivise the reduction or eventual phasing out of others. Guaranteeing a buyer will spur infrastructure to collect plastics and encourage brands to invest in new technologies to facilitate sorting, like artificial intelligence and digital watermarking, which involves using scannable codes carrying information about a package that can be detected by cameras on sorting lines.

The polluter pays principle could also be used for single-use products that are expensive or otherwise difficult for waste management authorities to deal with – like diapers, wet wipes, cigarettes and chewing gum. Relying just on cash-strapped governments, particularly in emerging markets, to magic up the money to collect waste and recyclables doesn't work. A whopping 2.7 billion people globally don't have access to waste collection.

Many consumer goods companies are coming round to the idea. They've historically pushed back against regulation that would saddle them with higher costs, arguing that they'll be forced to raise prices for already stretched consumers. The counterargument is that making the polluter pay is a fair system and that prices today, for many products, are artificially and unsustainably low, thanks to both oil subsidies and policies that allow companies to externalise environmental costs. Under the polluter pays model, the buyers of items like to-go food in disposable containers and disposable diapers shoulder the full cost of these items (for the latter, subsidies could be offered for families that genuinely need the financial help). It's also increasingly clear that without getting brands to shoulder the cost of their waste, infrastructure will remain drastically underfunded and we'll all, at best, just keep running in place.

Making the polluter pay, although raising costs for consumers at first, will inevitably motivate companies to innovate, shifting to less polluting alternatives that should rapidly decrease in price as volumes climb.

The alternative of not implementing the polluter pays principle, and not incentivising companies to consider the ramifications of the products they've spent billions of dollars hooking us all on, comes at a high price. Health problems linked to chemicals in plastics,[6] declining biodiversity and a warming planet are already costing us hugely.

*

At the back of the Carrefour store at which I met Bertrand Swiderski sits a large blue and white rectangular container with a hinged flap. On shelves next to it is a seemingly random collection of products. Bottles of Coke, wine and water sit side by side with jars of Nutella, beans, pureed vegetables and Nivea aftershave.

The collection has, in fact, been carefully curated. It's the product of an effort called Loop that aims to shift companies like Unilever, P&G, Coke, Pepsi and Nestlé to using returnable, refillable glass, plastic and metal containers for some of their best-known brands.

Loop was started by Tom Szaky, a shaggy-haired Princeton dropout who began his career in waste, turning worm poop into organic fertiliser, which he sold in old plastic drink bottles. Collecting the bottles inspired him to start a take-back programme for hard-to-recycle waste, like toothbrushes, mascara wands and cheese wrappers, which he turns into benches and garbage bags. Called TerraCycle, the firm charges companies a fee and, in exchange, they can put a label on their products touting them as recyclable through TerraCycle.

Szaky already had a line into many of the big brands and in 2019 he launched Loop with the aim of moving beyond recycling waste to

---

[6] A 2024 study led by Leo Trasande, a professor of paediatrics at New York University's Grossman School of Medicine, puts a $249 billion annual price tag on diseases attributable to plastics.

preventing it. He got Unilever and Pepsi to put their stick deodorants and breakfast cereals into refillable steel containers and P&G to put shampoo into a reusable aluminium bottle. Heinz agreed to refill glass ketchup bottles and Ferrero said it would refill glass Nutella jars. The effort got lots of publicity and the brands were happy to go along with it.

In France, the returns process is pretty simple. Consumers buy a large, closeable, reusable bag to put all their returns into. Then, rather than scanning each individual item, they scan the bag and drop it into the blue and white bin. Depending on the package, Loop or a third party cleans it and then the third party or the brand refills it so it's ready to be sent back to Carrefour.

I first wrote about Loop for the *Wall Street Journal* when Szaky launched the effort in January 2019. It sounded like a fun idea, but I had doubts about whether it would scale. Compared with disposables, the reusables were more inconvenient, offered people a very limited choice of products and were more expensive. When I checked in with Szaky towards the end of 2023, it was obvious that Loop was struggling to get traction. He explained that the returnable containers had been trialled by several large grocers, but only two had moved from the pilot phase into a wider rollout – Carrefour in France and Aeon in Japan.

During the hour I was at the Carrefour store, many customers came through the main doors but nobody walked over to the reusables section. Loop was offering a grand total of 40 different varieties of products, a small slice of what was on offer in the rest of the store. Bertrand Swiderski told me that Carrefour was rolling the reusables out to more stores because returnable packaging is something 'the consumer of tomorrow will demand'. The consumer of today, he admitted, has hardly any appetite to lug bulky bags full of used containers back to stores.

There are efforts afoot to make reuse easier. Bower Collective, a young Bristol-based online retailer, offers detergent, body wash and other cleaning products in refillable plastic pouches. It piggybacks off the UK's existing postal system, asking customers to mail several of the pouches back together once empty.

Flexible packaging that needs to be washed is unlikely to work, but Bower refills the pouches without washing them. It can do this because the pouch's spout includes a one-way valve that prevents air and contaminants from entering as it empties. The pouches aren't ideal since they're hard to recycle – once they've been used several times, Bower sends them to be turned into a plywood alternative – and they can't be used for food or for very large quantities. But compared to regular-sized single-use bottles of handwash and detergent, they do save on carbon emissions and cut down on waste. They could be a convenient solution for people who don't want to store bulky refillable containers.

In 2024, Ocado, an online UK supermarket, began selling rice and pasta in reusable containers. Once empty, the containers can be handed back to delivery drivers – who'd otherwise be returning to warehouses with empty vans – to be cleaned and refilled. Ocado has worked to make the packaging 'charmingly ugly' to discourage people from keeping it after it emerged that people were keeping their Loop containers.

Innovation to make reusables more convenient is much needed. But uptake for all these efforts is low. Szaky told me that the absence of regulation has been a big drag on Loop's growth – it's dependent on companies voluntarily being willing to absorb the added expense of investing in a new system. While the sustainability departments of all the retailers were interested, the commercial teams were entirely focused on the bottom line, he said.

Szaky's experience reminded me that although we talk about companies as monolithic entities with a single value system and a single way of doing things, they aren't really this way. There are factions within every company: what Coca-Cola wants to do is sometimes at odds with what its bottlers want, and what companies' environmental departments think would be a terrible package or a dubious claim is often one the marketing departments are most keen on. The folks who generate money almost always win.

In 2023, I learned of another reusables project Szaky is involved in. When he told me about it, I was so stunned I had to ask him to repeat himself. It turns out that P&G, after all its mudslinging at cloth, in 2020 – very quietly – acquired a cloth diaper brand called Charlie Banana. Even more crazily, P&G then hired Szaky to run a Loop-branded diaper washing and drying service for parents – the same model it drove virtually into extinction all those years ago.

Charlie Banana was founded by Gaëlle Wizenberg, a free-spirited woman originally from France who spent most of her childhood living on a boat in the Caribbean. Wizenberg dropped out of high school and later moved to Canada, where she worked for a software company and studied interior design, which she views as being largely about ensuring a place has the right feng shui. By 2006, she was living in Hong Kong and pregnant with her first child. She began looking for cloth diapers but found none in the city, so eventually settled for buying some second-hand ones on eBay.

The episode propelled her to create her own cloth diaper brand. She put her design skills to work in making the new diapers and wrestled her fluffy Persian cat into early versions. 'He didn't pee on them,' she assured me. 'It was more just modelling.'

Within a couple of years, Wizenberg had convinced Target, the big US retailer, to sell her brand online. Sales took off. Then, in 2019, a Geneva-based Pampers executive messaged Wizenberg on LinkedIn. 'He said he was in love with Charlie Banana,' she remembered. 'He had just had a baby and was using it.'

Although Wizenberg was willing to spend hours discussing Charlie Banana, she clammed up when I asked more about her interactions with P&G and the company's motivations for buying her brand. She said P&G had made her sign a nondisclosure agreement and she was nervous about breaking it. 'Basically, they realised the cloth diaper category was growing and they wanted a brand that was well established,' she said cautiously. 'We were a very small company, but we were in all the big retailers.'

Wizenberg put me in touch with the P&G executive who had first contacted her, and he in turn sent my email to Andy Daly, P&G's brand director for Pampers. 'More than happy to take any questions,' Daly wrote to me. 'Let me just double check with our internal communications folks to get their blessing and we can set up a date.'

The department did not give its blessing and Daly and I never got to speak.

Instead, a P&G spokeswoman sent me an email listing the benefits of disposable diapers. 'Over the years, disposable diapers have become essential goods for generations of babies & parents, protecting, supporting and empowering them,' it began. She went on to explain how P&G's disposable diapers have become more eco-friendly over the years through lightweighting[7] and

---

[7] P&G says it has cut the weight of its disposable diapers by 40 per cent over 25 years in the US. Its learnings from piloting Charlie Banana include that the majority of parents said switching to cloth would be difficult or impossible, that 67 per cent of those washing cloth diapers in the US and UK used high-temperature washes, and that 47 per cent used dryers. Cloth diaper wearers used more cream and baby powder. P&G offered no positive learnings from its trial with cloth diapers. However, a study published by the UK government in 2023 – 'Life Cycle Assessment of Disposable and Reusable Nappies in the UK 2023' – concluded that the emissions from cloth diapers are substantially lower than those from disposables.

incorporating more plant-based material. She said P&G had acquired Charlie Banana 'to offer consumers a choice and also to learn parents' habits and practices with cloth diapers'.

When I emailed Szaky in April 2024 to ask how Charlie Banana was doing and follow up on the Loop Diaper projects' progress, he sounded deflated. Since the initiative had kicked off in 2022, it had had only 300 users, he disclosed. 'Consumers still love the convenience of disposable diapers,' he wrote. 'Honestly, Loop Diaper is struggling to find adoption among consumers.'

As with Szaky's other Loop effort, the learning here is clear. In instances where reusables are the most environmentally beneficial option, we need regulation to level the playing field against disposables. Such regulation could require a growing percentage of what a company sells to be reusable or in reusable containers. It could slap taxes on disposables or, under the polluter pays system, charge companies that choose to use disposable packaging higher fees than those that choose reusables.

Think of tiered fees like car insurance – just as reckless driving can lead to higher insurance premiums, environmentally harmful products or packaging results in higher fees. Using tiered fees should encourage companies to 'drive safely' by using more sustainable materials, ultimately reducing their financial burden and benefiting the environment and human health.

Such fees can also push industries to innovate and spend some of their marketing dollars on ensuring less environmentally damaging products take off. Many start-ups are selling laundry detergent pods and sheets in small cardboard boxes, hoping to woo consumers away from heavy detergent bottles that contain mostly water and are made from plastic. Although big companies like Unilever and P&G have followed suit, the multinationals have chosen to prioritise their existing bottle detergents when it comes to marketing and investment.

Beyond this, pay-as-you-throw schemes – which charge people variable rates based on how much rubbish (not recycling) they throw away – along with higher landfill fees and incinerator taxes, would encourage us all to reuse and recycle more.

Bill Rathje, the garbologist, discovered that the larger the receptacles people have available for trash, the more trash they produce. He called this phenomenon 'Parkinson's Law of Garbage' after the 1950s observation from historian Cyril Parkinson that work expands to fill the time available for its completion.

Charging people according to how much they discard makes them aware of how much they're discarding, what's filling up their bins and whether they can buy less of this – or use a different version. Without such measures to incentivise change, most of us will continue to buy stuff and throw it away on autopilot.

*

In April 2024, I flew to Aarhus in Denmark to see the city's new returnable cup programme. Located along a bay to the west of Copenhagen, Denmark's second-biggest city is home to a student-heavy population of some 350,000 people. When I hopped off the airport bus in the city centre, it was a sunny day and several people walked by clutching durable white cups with blue lids and the word REUSABLE written prominently on them.

For years, chains like Starbucks and McDonald's have trialled giving people drinks in reusable cups at a handful of outlets in the US, UK and other countries. These programmes attract big publicity, but don't expand and usually close down entirely after a few months.

Coffee shops have also – unsuccessfully – long offered consumers discounts for bringing their own cups. For most people, shaving off a little money doesn't outweigh the inconvenience of toting around a used coffee cup, dregs and all, for the rest of the day. For the few who do bring in cups, coffee shops often aren't set up to handle these. Baristas accepting the reusables have been spotted making coffee in disposable cups and decanting this into consumers' reusable cups before throwing the disposables away. The issue is that the reusables are too tall for their machines. At the 2024 Olympics in Paris, bartenders poured Coke from small disposable plastic bottles into reusable plastic cups, an exercise that created more waste than just handing out plastic bottles.

The Aarhus programme's originator is Simon Rossau, a curly-haired 32-year-old who works on sustainability for the city's waste department and lives on a houseboat. Aarhus was spending over $4 million a year on managing waste in its streets, parks and waterways when it hired Rossau in early 2022 to help slash the amount of waste the public discarded. He commissioned an analysis showing that 48 per cent of all waste binned or littered in the city's public areas was single-use packaging. Disposable cups were among the worst culprits; the city was fishing 100,000 of these out of the bay each year.

By January 2024, Rossau had a reusable cup programme going. Tomra, a large maker of reverse vending machines and sorting equipment for recyclers, had provided the machines and set up a cup washing, drying and returns service in the hopes of convincing other cities that if Aarhus could do it, they could too.

The Aarhus programme has several elements that make it a cut above others. All the participating venues use the same containers, taking a reusable cup triggers a deposit that strongly encourages people to return their containers, and returns are quick and easy. There are six versions of the cups. White ones are meant for hot drinks, while taller frosted cups are for juice and soft drinks. The cups are available in small, medium and large.

Two and a half months into the programme, when I visited Aarhus, Rossau had convinced more than 50 cafes – and a handful of other venues, like food markets and concert halls – to use the standardised cups, which could be returned to 28 vending machines dotted around the city. The return rate was 86 per cent and Rossau was hoping to expand the trial to include containers for common takeaway foods like salads and sushi.

He guided me towards Aarhus Street Food, where I bought some tacos for lunch and a Pepsi in a frosted reusable cup. Returning my used cup was a doddle. I tapped my card against a reader on a reverse vending machine, put the cup into a hole in the front and a message flashed up saying the 5 kroner deposit would be back in my account within two days.

Using a container as many times as possible reduces the emissions tied to its disposal and production. In Aarhus, the deposit and the convenience have both helped keep the return rate up. It also helps that Danes are used to making returns: Denmark has a successful container deposit programme that sees plastic drinks bottles returned for recycling or refill at a rate of 91 per cent.

The cultural element is less talked about but is a huge driver of how people behave. It's often underpinned by regulation. During my time in both Aarhus and Copenhagen, I watched as pedestrians consistently waited patiently at traffic lights, even on the smallest streets that were empty of cars. A Danish friend explained that fines for breaking the rules are hefty and enforcement is strict.

Standardised packaging that can easily be returned to a large number of conveniently located machines, washed, dried and sent back to local restaurants or cafes can help lower emissions. The programmes are also simpler, cheaper and more convenient than 'island systems' in which consumers must return to individual chains to drop off their specific cups.

But the Aarhus programme is a voluntary one and isn't backed by enforcement. Cafes don't have to offer people drinks in reusable cups and don't have to hit any reuse targets. The 86 per cent

return rate drops to just 60 per cent once Aarhus Street Food –
easily the programme's biggest cup generator – is taken out of
the equation. The city needs more reverse vending machines, but
Rossau has had pushback from many venues – including the main
train station – whose operators don't want the machines on their
premises, and from McDonald's, which has declined to participate in the programme.

It's also clear that the deposit should be higher. Deposits that
are high enough and move in lockstep with inflation are a great way
to incentivise people to return packaging for recycling or reuse. It
all comes down to putting an appropriate value on used packaging.
You don't see dollar bills or even 10 cent coins on the street. As long
as used packaging has a high enough value, it's much less likely to
be binned, littered or uselessly stockpiled at home. To avoid over-
burdening consumers, some reuse models are exploring triggering a
charge only if containers aren't returned.

But Rossau, reliant on the goodwill of cafes that are voluntary
participants, worries that raising the deposit will alienate customers.
'We need legislation to make single-use as expensive as reusables,'
he said. For that, Aarhus is dependent on legislators in Denmark's
capital, none of whom, as of April 2024, had come to see the programme for themselves. 'We always say the trip from Copenhagen
to Aarhus is much longer than from Aarhus to Copenhagen,'
Rossau said wryly.

*

In November 2023, I met with McDonald's head European lobbyist, Serge Thines, at a large McDonald's on Boulevard de la
Madeleine in downtown Paris.

While McDonald's did not sign up to the Aarhus programme,
proposals to mandate reuse had been gathering steam across

Europe, posing a threat to McDonald's ultra-fast, ultra-cheap business model. Since 2023, McDonald's outlets in Germany have been forced to abide by a law requiring that they offer reusable cups to customers, including for takeaway and home delivery orders.

A towering man who speaks five languages, Thines was eager to tell me how disastrous reusables are for the environment. He confided that just 2 per cent of customers in Germany chose the reusable cups and that only 40 per cent of these were returned, despite a hefty 2-euro deposit. 'The idea is that if you have a deposit return scheme, then customers will bring back the reusables,' he said. 'In Germany, we see that that's actually not the case.'

At the time, I was bemused by how poor the uptake and return rates were. Then a few months later, while visiting family in Germany, it became clear why this was. At a bustling McDonald's near the central square in Munich's old town, there were no signs anywhere in the restaurant – and nothing on the menus behind the counter – explaining that people could ask for a reusable cup.[8]

Despite the fact that many Germans still pay using cash, the only way to order a coffee in a reusable cup was through digital screens – which I did. Upon collecting my cappuccino from the counter, I was given a yellow plastic cup that offered no indication it was reusable. 'Where do I return this?' I asked the staff member who handed it to me. 'I don't know,' she shrugged, waving for me to keep moving.

I eventually figured out that to get my deposit back I'd need to queue behind everyone ordering at another counter. Although I had paid by card, when I got to the front, I was handed my deposit in coins with no option to tap for a refund. Coming from a company

---

[8] I can't read German so I asked my German cousin (thanks, Sarah!) to come along to be my second set of eyes and also ensure I wasn't missing any signs in German advertising the reusables.

that has speed and convenience in its DNA, the experience felt particularly inconvenient.

In France, since 2023 McDonald's has been forced to serve nearly all its food and drinks in reusable containers to people dining in.[9] When I met with Thines in November, the French law appeared to be the tip of the iceberg. In the coming weeks, European regulators were slated to vote on similar measures for fast-food chains across the bloc. Thines and other McDonald's representatives lobbied ferociously – and ultimately successfully – to water down the European proposal. While environmental campaigners were calling for mandatory reuse targets, the law that ultimately passed had none.

McDonald's argued, among other things, that reusables raise emissions. 'We don't want France to be the template because we strongly believe it doesn't work,' Thines said as he showed me stacks of reusable nugget and French fry containers in McDonald's kitchen. 'We are for recycling,' he added.

Despite Thines's claims, my reporting on McDonald's recycling efforts has shown that the chain is still struggling with all the same problems it had convincing consumers to separate out waste as it was in 1990. 'Even though people come in and they eat with their hands – there are no knives and forks or anything – when they go to the bins, they don't want to touch their food,' explained Helen McFarlane, sustainability manager for McDonald's in the UK. Several years ago, the chain's research showed that customers simply bundled everything into the paper bag their orders came in so they could throw it all in a single bin. Asking them to separate pickles and soggy teabags from paper containers landed badly. 'People were quite angry about having to do it,' McFarlane told me.

---

[9] France has made an exception to the reusables rule for burgers and sandwiches, which can still be sold in disposable packages so long as these are wraps rather than boxes, and for condiments like ketchup.

Recyclers don't want paper packaging with leftover food in it, and today, many McDonald's restaurants don't even provide recycling bins for paper. As I mentioned previously, the chain swapped its foam cups for plastic-lined paper ones that are very hard to recycle. The upshot is that most of its packaging the world over hits landfills or incinerators if it isn't littered.

Despite this, my knee-jerk reaction when I first heard about the French law was to concur with McDonald's. Producing durable plastic containers and washing and drying them generates large greenhouse gas emissions – surely this wasn't better for the environment? McDonald's sent me a study it had commissioned showing that packaging waste, emissions, plastic, water and energy use would all rise if Europe were to shift to reusable containers at scale.

But the more I researched the issue, the more I realised it wasn't as open-and-shut as McDonald's was implying. For one thing, reusables are being compared against disposable packaging that's had decades to become highly efficient and cheap. Remember how cellophane in the 1920s was stored in a safe because it cost $2.65 a pound? Well, as it scaled, costs plummeted. Within two decades, the price stood at 33 cents a pound.

I had also since seen counter-studies commissioned by waste-reduction non-profits which found that reusables can cut waste and emissions. As I described in Chapter 5, when it comes to life cycle analyses, the assumptions companies choose – for metrics like return rates, the use of renewable energy and the temperature at which reusables are washed – have a big impact on the outcome.

Anything new has teething pains, but in time companies learn to optimise and make things more efficient. Reducing transport distances, using renewable energy, choosing low-carbon materials and ones that avoid dangerous chemicals, ensuring containers are returned and reused at a high rate, and recycling them at the end of their lives can all help mitigate the environmental impacts of reusables.

When I met Thines, McDonald's had been using reusable containers in France for several months and had already done quite a lot of optimising. It had added radio frequency identification tags so it could find containers thrown into the wrong bins. It had increased the size of bin openings to make it clear where consumers should put the returnable containers once they were done with them. It was using scanning guns that could count and itemise hundreds of containers in seconds. It had a powerful new dishwasher that took just two minutes per load, and a dryer that worked in the same amount of time. It had switched to ribbed reusable containers from smooth ones that scuffed easily.

The containers are made from a durable plastic called Tritan that's touted as being free of bisphenols. On social media, French diners tweeted about how much they loved McDonald's new containers. Elium, the French firm that designed them, was eager to speak with me, but McDonald's wouldn't allow it. A spokeswoman told me McDonald's didn't want to call any more attention to the containers since it didn't support the idea of reusables.

There are challenges in abiding by France's law, and I can see how small restaurants would need to carefully redesign their storage space to make room for bulkier reusable containers. However, it's also possible that the law – if other countries were to adopt something similar – could push McDonald's to design new containers specifically to be reused. Right now, it's just adapted its disposables: the French-fry containers, for instance, are bulky versions of the disposable sleeves when there's no reason why fries couldn't be served on a slim side-plate.

Brands have poured many millions of dollars into hooking us on disposable products and shaping our consumption habits. Remember all the work P&G did to convince new mothers that cloth diapers were old-fashioned? Or the intensive marketing Unilever did to get Indians to ditch homemade shampoo in favour of Sunsilk? Or how Coke and Pepsi switched to disposable containers even though people liked returnable glass? Consumers weren't clamouring for disposable packaging – it was the big brands who convinced us we needed it.

Imagine if these same brands turned some of their money and ingenuity towards bringing reusables into the 21st century. Of course, it's a transition that will need to be done carefully. Reusables likely won't ever make sense for many everyday products – things like salad leaves, potato chips, small portions of biscuits and chocolate bars. They may not work outside of densely populated cities. And if reusable systems are badly designed or executed, emissions, water usage and waste could all rise.

Where we do use reusables, what material they should be made from is a million-dollar question. All materials come with environmental footprints and are only suitable for certain types of things. Glass is made in furnaces at very high temperatures – its production creates 95 million tonnes of carbon dioxide globally each year. It's also heavy to transport, leading to increased fuel use, and breaks easily, which can harm paws, feet and hands as well as lead to waste. Aluminium is mined from bauxite, which has a range of negative environmental impacts, one of which is deforestation. When used for food and drink, aluminium is usually lined with plastic that can contain hazardous chemicals like bisphenols.

Plastics are light, unbreakable and can have lower carbon emissions than other materials. But given concerns about microplastics and chemicals that leach from plastics into food and drink, there are big questions overhanging even durable plastics. These also encompass many bio-based plastics, which have been found to contain some of the same problematic chemicals as conventional plastic and can compete for land used to grow food. Bio-based plastics – made from materials like sugarcane and starch rather than fossil fuels – aren't always readily compostable or recyclable and can linger in nature for years if littered.

Picking materials carefully – and ensuring they can be reused repeatedly, don't use dangerous chemicals and can be recycled at the end of their lives – is key to minimising any unintended consequences from a shift away from conventional single-use plastics.

Still, back in 2023, by the time I was done with my tour of the backrooms of the McDonald's restaurant in Paris, I felt more upbeat about the prospects for reusables than I had when I stepped through the door.

I phoned Shelby Yastrow – the former McDonald's general counsel and one-time defender of its clamshell containers – to tell him about the Paris effort. 'I'm not surprised they're able to adapt to it,' he said. 'The company is incredibly good at getting things done. I can give you a million stories about how we put in drive-thrus and put in one restaurant every day.'

Yastrow remembered a similar law to France's that was proposed in Germany during his time. McDonald's successfully killed it by arguing that reusable containers used more emissions, were more polluting and were harmful to human health due to bacteria. 'There's a lot of ways to prove a lot of things,' he told me. 'You can get studies to show anything, I guess.'

When I asked Yastrow if he'd make the same arguments today, given the amount of waste McDonald's is still producing and how widely littered its containers are, he paused. 'Even though I would have guessed McDonald's could do what you saw in Paris, at the time I would have resisted it and tried to keep the polystyrene,' he said. His main objection, he added, was if Germany did it, the Netherlands would likely follow suit and then other countries. The fast-food chain could be on the hook for billions of dollars in extra costs.

'That would have been enough of a reason for me to resist,' he concluded. 'I was an employee of McDonald's and we were in the business of making money if we could.'

CONCLUSION: WHAT CAN I DO?

Among the stories Shelby Yastrow told me was one about the uproar he caused when he wanted McDonald's to shift from using bleached to unbleached paper for its bags. 'You'd think I was asking them to burn down churches,' he said.

Once he convinced the packaging folks at McDonald's, he ran into a new hurdle: the fast-food chain's suppliers claimed they couldn't make the switch. When McDonald's eventually started sourcing the bags from a new supplier, suddenly the others all said making unbleached bags was no problem.

Large consumer goods companies are the conduit between the thousands of companies who make packaging, chemicals and plastic resin, and us – the end buyers of these products. The consumer goods companies have unrivalled power to make changes to the products we eat, drink, shower and clean with, and otherwise use on a daily basis. I chose to focus this book mainly on these companies because I believe they're the ones who have the biggest responsibility, and are in the best position, to tackle the problems that have emerged as a result of their – and by extension our – addiction to disposability.

As consumers, we in turn have an enormous opportunity to influence the shape of the future – by encouraging companies to change their products, packaging and marketing, and even to shift entire business models – for the obvious reason that they need our custom. If we're unhappy, that's not good for business. This gives us power – we just need to start harnessing it.

Is a product overpackaged? Does it use packaging unnecessarily? Is a company marketing its wipes or cups as biodegradable when there's no composting facility that will accept these? Is a plastic pouch labelled as 'recycle' or 'widely recycled' when it's not accepted by your local council or municipality? Has a company said it would stop using plastic packaging but continued to use it? Is it claiming its single-use, hard-to-recycle carton is 'eco-friendly'?

If a product you consume regularly makes you feel guilty, frustrated or angry, write to the company who makes it, telling them how you feel and what you'd like to see instead. Packaging is as much a marketing tool as a way to protect and deliver a product, and companies have a wide array of choices about what they use.

You can get your kids or your colleagues involved, including by creating a petition. This sounds like a big step, but it doesn't have to be. In 2018, ten-year-old Ella and eight-year-old Caitlin started a petition to get McDonald's in the UK to stop handing out plastic toys with their Happy Meals. They got so many signatures that McDonald's scrapped the toys. The company still hands out soft toys made from synthetic materials (in other words plastic) without asking if customers want them, and it's time someone started a new petition, but the point is that Ella and Caitlin got one of the most powerful companies in the world to make a significant change.

As you look to cut back on the amount you're throwing away, remember that a reusable container is only more environmentally friendly than a disposable option if, over the course of its life, it 'earns back' the emissions embodied in making it. So when you buy a cloth bag or a stainless-steel water bottle, try to own just one (or the minimum number you need), use it regularly and wash it in cold water.

Vote with your wallet – if you don't approve of what a company is selling, look for an alternative. Hand back stuff you don't want: my bugbears include straws, condiment sachets, pens and other conference tat, and carrier bags. Before you buy something, ask yourself 'Do I really need this?' It helps me to remember that many of my 'needs' have been created by big companies that pour millions of dollars into ensuring I'm never satisfied.

Before you throw something away, ask if you can repair, donate or safely reuse it. Pause for a moment and think about where it will go and what will become of it. I've found becoming mindful about what I'm consuming, where it comes from, whether I need it and what its afterlife might look like can motivate me to make more considered choices.

Admittedly the stuff I'm suggesting is pretty small potatoes in the grand scheme of things. Individual changes won't be enough to make a big dent in our plastic waste problem. But hopefully by now you have a good sense of how we got here and what needs to change to break out of the cycle of stagnancy and failure. The key lies in policy measures that rewire the skewed economics underpinning single-use plastics, and push companies to adopt new business models and reinvent supply chains. That being said, individuals can spark cultural change, which in turn motivates both regulators and companies to act.

These days, at meetings I go to in London, I've noticed that most companies have stopped putting out individual plastic bottles of water and instead offer water in reusable glass jugs and glasses. This cultural shift didn't happen overnight, but consumers protested loud and long enough that a company choosing to unnecessarily opt for single-use plastic bottles stands out – and not in a good way.

Conclusion: What Can I Do?

To reset the linear system within which we've all operated for the past 80 years, we need to call for consistent, clear regulation that's powerful enough to effect change and that is actually enforced. I've outlined some of what's needed to get reusables off the ground already and, in the FAQs that follow, I'll offer a few more suggestions about how regulation can improve recycling, which I believe does have a place but needs to be fixed.

Without regulation that forces whole industries to move together, companies are very reluctant to do anything that could lessen the appeal of their brands versus those of rivals – or to incur added cost and disruption. If they're publicly traded, the contorted system that binds them to maximising profit for shareholders over anything else can also effectively mean their hands are tied. Left to their own devices, many companies have little choice but to take the easy way out.

FAQS

Is single-use plastic bad?

It depends on the perspective you're taking. On the one hand, using plastic can save on carbon emissions compared to other materials, and plastic offers clear functional benefits, including helping to keep perishable products fresh for longer, reducing food waste.

On the other hand, single-use plastics – which make up about half of plastics thrown away – are largely derived from climate-warming fossil fuels and are projected to play a bigger role in keeping the oil industry booming as the world turns to using more renewable energy. The majority of single-use plastics today are technologically difficult or uneconomical to recycle, and often end up being littered, dumped or escaping into the environment. Even if plastics are properly managed at the end of their lives, during use they can leach chemicals and microplastics into the food and drinks we're consuming at levels many scientists believe aren't safe.

Rather than giving companies free rein to use single-use plastics mindlessly because they're so cheap and ubiquitous (I recently received a delivery of a plastic-wrapped T-shirt still on its plastic hanger in a larger plastic bag), regulators should put in place laws that require companies to use plastics far more carefully for a more limited range of applications for which they are truly the best choice.

Should I try to use less single-use plastic?

Given that all materials come with an environmental footprint, the best course of action is to try to use fewer unnecessary items of all kinds – not just less plastic – and create less waste overall.

If you're faced with the choice of replacing plastic with another material, ask yourself some of these questions: Is the alternative you're switching to better for the environment? If so, why? What is it made of? How heavy is it? Can it be easily recycled? Can it be composted and will it actually end up in a composting plant? Can it be reused and, if so, will you actually reuse it many times? Is switching away from plastic likely to be better for your health? Are there unintended consequences to switching away, and if so, can you mitigate these?

Some everyday ways in which I cut down on packaging include choosing unwrapped produce at supermarkets where possible and looking up the best ways to store herbs, salad leaves, fruits and other fresh food at home. (Did you know that loose apples stored in the fridge below 5 degrees Celsius can last 102 days? That's 70 days longer than if you store them at room temperature in the UK.) I also buy the amount I know I'll eat, keep tabs on what's likely to go off and use it up before it does.

My family and I don't regularly get takeout, but when we do, I try to remember to ask restaurants to leave out the single-use cutlery, napkins, straws and condiment sachets. If I'm collecting food, some places will allow me to provide my own reusable container. I usually bring my lunch to work in a reusable glass or stainless-steel container. When I know I'm going to grab food to go, I personally find it easy enough to carry around a reusable water bottle and to tuck a handkerchief and spoon into my bag. Adding coffee cups to this mix is admittedly harder, which is why I think an Aarhus-style system that lets me conveniently return standardised reusable cups all over London – or, better yet, across the whole of the UK – would be amazing.

I favour using bars of soap over body gel and buy large recyclable containers of cooking oil, vinegar and hand soap, which I decant into smaller reusable containers. Menstrual cups are an easy swap for pads and tampons; they function well and are substantially cheaper over the long run. I don't line individual trash cans with plastic, instead decanting these into a single plastic bag that we hand over to the garbage collector once every few weeks. I scoop and flush my dog's poop rather than bagging it when she goes in our garden.

I also reuse stuff where I can, turning large glass olive jars into food storage containers and plastic bags into packaging for parcels. I used to reuse plastic takeout food containers but do so less nowadays, because I don't feel confident that these are designed to withstand degradation through multiple uses. I've asked my kids' nursery to stop returning soiled clothes in single-use plastic bags and instead put them in a reusable one I've provided that lasts many months.

Beyond packaging, I'm lucky enough to live somewhere in London that sends our food waste to an anaerobic digestion facility. I minimise what I put in food waste by saving vegetable stems and incorporating these into my dog's food. I serve my picky toddlers small portions and give them more as they ask for it. I drop off plastic film for recycling at our large local Sainsbury's supermarket. While there are regular media reports about how flexible plastics get incinerated or sent overseas, dropping these off at least gives them a chance of being recycled, while putting them in general waste means they'll definitely be incinerated or landfilled.

As a family, our biggest source of waste is disposable diapers. We only use washable wipes at home, but there's no denying how bulky that diaper trash bag is at the end of each week. If a diaper washing service was locally available at an affordable price, I'd spring for one, if only to assuage the guilt.

I buy lots of stuff second-hand, sell or pass on clothes, toys and household items we aren't using (Facebook Marketplace and our online neighbourhood 'freebies' group are both great for this) and check if we can repair appliances before dropping them off for recycling at a collection point. A few months ago, I realised that buying little things online had inadvertently become a tool I was using to destress. Acknowledging this and taking steps to mitigate it, like plugging my phone in away from where I wind down, has been useful. Overall, I just try to be conscious of how much waste I'm producing. I still throw away a lot, but I buy less – and waste less – than I used to.

Is recycling a scam?

The reason so many people think recycling is a scam is that it's been oversold to us: companies have worked hard to give the impression that plastics are widely recycled when in practice only a small slice – mainly rigid PET and HDPE containers – are.

Recycling, despite all its problems, can be worthwhile. It can save resources that would otherwise go into the extraction of new raw materials, often the most energy-intensive part of making a new product. It can also keep materials out of landfills and incinerators, reducing emissions. Whether something should be recycled comes down to whether doing so results in a net-emissions saving – transporting recyclables a long way by truck or using energy-intensive recycling technologies may produce more emissions than using virgin material.

Why are plastics recycling rates so low?

Recycling is a business. It only works if a) there are buyers for recycled plastic, and b) these buyers are willing to pay more than it costs to recycle.[1]

Today, there are problems with both these things. The cost of collecting, sorting, cleaning, trucking and pelletising used plastics is often far higher than buying virgin plastics, and the quality is lower. Unsurprisingly, that translates into a lack of willing buyers.

Of the plastics that are recycled, most are downcycled into lower-quality products like carpets or plant pots that can't be further recycled. One reason for this is it's too expensive to sort plastics by colour. Another is that turning used food packaging into new food packaging requires extra cleaning, which is also expensive. Plus, many packages incorporate multiple kinds of plastics or other materials and so can't be recycled at all using mainstream recycling technologies.

Regulation that requires companies to bear the cost of recycling their products, bans some non-recyclable plastics, requires packages be designed for recycling, charges companies higher fees for using pigments and other additives that narrow end markets, levies deposits on common to-go containers, limits the plastics and additives that companies are allowed to use, standardises containers to create high volumes (which helps end markets develop) and requires packaging to be made from a minimum amount of recycled material could help address some of the big problems that have long plagued plastics recycling.

---

[1] This applies to all materials. In the US, the recycling rate for glass containers is just 31 per cent and for aluminium packaging it's 35 per cent, according to the Environmental Protection Agency. That's better than the 13.6 per cent recycling rate for plastic packaging, but it's still pretty terrible.

FAQs

What can I do to support recycling?

You can help on both the supply and the demand side. To increase demand for recycled plastic, buy recycled packaging where you can. Write to brand owners telling them you'd like them to use more recycled material. You could also choose packaging that's designed to be easy to recycle. Where possible, I favour clear plastic over coloured since the latter will inevitably be downcycled.

On the supply side, pay attention to what you're putting in the recycling bin. If you're unsure whether something is recyclable, check your local council or municipality's website and if you're still unsure, leave it out.

A few rules of thumb: glass, plastic and aluminium drinks containers, along with newspapers and cardboard, are thrown away in high volumes, which makes them widely accepted for recycling.

Scrape all food out and rinse minimally with cold water. Remember that recycling is an emissions game, so if you're scrubbing containers with hot water and soap you're lowering – possibly even negating – the net emissions saved from recycling. For the same reason, try not to drive to places just to drop off recyclables. Instead, save them up and drop them off on a trip you'd need to make in that direction anyway.

If you're out at a park or on the go, try to bring drinks bottles and other recyclables home to put in your kerbside bin. Street-corner recycling bins are often so highly contaminated with cigarette butts, chewing gum and food that entire loads are rejected. Squash down drinks containers and then put caps back on so they get recycled too. Small non-metal items – a good rule of thumb is less than two inches by two inches – usually can't be sorted for recycling, so put those in general waste.

Don't put plastic bags or other flexible plastics in kerbside bins, since they gum up machinery if not collected separately. Bags can usually be dropped off at supermarket collection points for recycling.

Whether they're recycled in practice depends on there being buyers in place – this could be facilitated by the regulation I outlined previously. Companies are moving towards simpler multilayer packaging, made from plastics that are compatible for recycling, so in time you should be able to drop off a wider range of flexible plastics (or have these collected kerbside as funding from brands under the polluter pays model expands collections).

Multi-material items – like soap top pumps that use metal springs, plastic-lined paper envelopes, blister packs for medicines, and many sachets – can't be easily recycled, so leave these out too.

Many councils today don't accept juice and milk cartons in kerbside bins because they're made from a mix of paper, plastic and aluminium. Minimise your use of these packages, but for the ones you do buy, look online for drop-off programmes.

What's greener: cloth diapers or disposables?

This is an ongoing question that nobody neutral has a great answer to.

Most of the studies that exist are funded by cloth or disposable diaper makers. However, in 2023, the UK government published a study on diapers. It found that disposable diapers have a higher global warming potential while reusable diapers have a higher environmental impact in areas like marine ecotoxicity, in part because of the electricity used to wash and dry them and the water used by washing machines. The results are specific to the UK, where disposable diapers are far more likely to be incinerated rather than landfilled, the grid has become less carbon intensive and washing machines have become more efficient.

The way I see it, the answer depends on where in the world you are. Is water scarce? How much renewable energy does your grid use? Are the diapers likely to be dumped or come into contact with waste pickers? Can you dry cloth diapers outdoors or will you end up using a tumble dryer? Do you have access to a diaper service that centrally washes and dries? Cost and convenience will undoubtedly play a role too.

Diaper recycling or composting may also yet become a reality, if the economics align. In France, private-label diaper maker Celluloses de Brocéliande has developed a fully compostable diaper (including compostable super absorbers) made from bamboo viscose, cornstarch and corn cellulose. Unsurprisingly, there's still work to be done: the diaper isn't as absorbent as regular disposables are while the resulting compost can't be used to grow food, meaning it doesn't sell for a lot. Eric Vilmen – who heads product development for Celluloses de Brocéliande – is working to convince regulators that high-temperature composting kills all pathogens. 'If we only use it for flowers and parks it doesn't make sense,' he says. 'We need to increase the potential of using it more widely.'

In Wales, a diaper recycling project called NappiCycle is turning used diapers into material to make roads and decking. The project hopes to avoid the pitfalls faced by P&G by selling a higher margin recycled pellet rather than plastic film and cellulose individually. NappiCycle founder Rob Poyer is betting that rising taxes on landfills and incinerator emissions will make diaper recycling more economically viable in the next few years.

The best thing to do, of course, is potty train your child early. There's a method called 'elimination communication' which is based on the underlying philosophy that babies, like other animals, instinctively prefer to avoid soiling themselves. If parents can recognise a baby's signals that she needs to relieve herself – similar to tuning in to sleep and hunger cues – or can take an educated guess based on a baby's usual timings, they can hold her over a potty or sink. It helps to pair this with a sound – like 'psss' – which over time triggers a baby to go.

Christine Gross-Loh, author of *The Diaper-Free Baby*, says parents can start elimination communication with babies who are just hours old. The key, she tells me, is to help babies retain an awareness of their bodily functions. She advises not being a perfectionist and accepting that there will be misses. 'The best way to think about it is this is another alternative,' she says. 'No matter how much or little I do it, my baby will not think diapers are the only place to go.'

How can I protect myself against greenwashing?

Be sceptical about green claims, pause to think about the whole picture – including both waste and emissions – and try to connect the dots to the end. Paper isn't always better than plastic, biodegradable plastic often goes to landfills or incinerators where it doesn't biodegrade, and while a pricey bottle or shoe made from 'ocean plastic' sounds impressive, it does little to address the root causes of our plastics problems.

Ask yourself: Why is a company making this claim? Is it specific or does the company portray itself as 'eco-friendly' without explaining how? Does a product or package do what it claims? If so, is that good for the environment? And what are the trade-offs?

If you think the company's green claims are deceptive, you can complain to the Competition and Markets Authority in the UK – email them at *general.enquiries@cma.gov.uk* or use the online form. In the US, you can complain to the Federal Trade Commission using its 'report fraud' link. If you're a shareholder of a company, you can also complain to the Securities and Exchange Commission, which works to ensure that shareholder disclosures are accurate.

Isn't paper better than plastic?

It can be but it isn't always. Rather than a wholesale shift from plastic to paper, it makes sense to look at each product on a case-by-case basis that attempts to balance waste, emissions and likely health hazards to arrive at the best choice.

Paper is derived from a renewable resource and will break down when littered, but it still has a big environmental footprint. It can cause deforestation and harm an area's biodiversity. It's highly water-intensive, as well as chemical-intensive: a 2021 report found that 256 substances of concern – chemical substances that may pose risks to human health or the environment – can intentionally be used in paper and board packaging.

When it comes to holding food and drinks, paper isn't naturally good at repelling grease or moisture. It needs coatings, which are often made from plastic. Paper food packaging has also been found to use PFAs, the notorious 'forever' chemicals. Companies say they're phasing out the problematic PFAs, but researchers have raised concerns that replacement chemicals that function similarly could be harmful to our health too.

Food waste and non-fibre components – such as coatings – that make up more than 5 per cent of a package can mess with the economics of the paper recycling process. Many countries mandate that paper packaging used for food be made from virgin paper, which means recycled food containers can't be used to make new food containers, but must be downcycled into lower-grade material.

That being said, I do think paper that doesn't use dangerous chemicals could be a good choice in a few areas. One is for disposable packaging that's widely littered – including snack packets, bottle labels, chocolate and energy bar wrappers – and too small to charge a deposit on. Using virgin paper isn't inherently bad as wood can be grown and harvested as a crop.

Two caveats here: the coated paper will likely keep its contents fresh for a shorter amount of time than plastic, so an unintended consequence could be increased food waste if supply chains and consumer communication aren't well managed. And if the coating isn't biodegradable, the package could leave behind long-lasting microplastics if littered.

Paper could also play more of a role in packaging dry products. More companies are stripping water out of laundry detergents, instead shipping concentrated dry pods and sheets in paper boxes. If you stop and think about it, it's mad to ship a product that's mostly water around the world if you can just ship the concentrate. There's potential to sell dry versions of more products, but companies would need to invest the same millions convincing consumers to switch to these as they did hooking them on liquid versions in the first place.

Are reusables the answer to our plastic waste problem?

They're certainly *an* answer but they won't work for everything.

From an emissions and cost point of view, reusables can make sense when replacing some rigid packaging, but supermarkets are unlikely to be selling you potato chips, chocolate bars and salad leaves in reusable containers anytime soon.

Key to keeping emissions and costs down is scale and high return rates – these could be achieved through mandatory reuse targets and deposits that keep pace with inflation. Standardised reusable packaging shared across brands for products like soft drinks, yogurt and shampoo could cut carbon emissions, complexity and costs, since containers can be returned from drop-off points to the nearest factory for washing and refilling, cutting down on travel distances. As I described in Chapter 13, supermarkets making deliveries of online grocery orders have begun experimenting with taking back used containers for refill in vans that would otherwise be returning empty – a modern-day version of the old milkman model.

So far, consumer goods companies that have waded into reusables on their own – for instance, with brand-specific refillable deodorant sticks and cleaning sprays – have seen little success. Uptake is also low at grocery stores that allow people to bring their own containers to refill a limited range of dry goods like rice and pasta. Consumers need to plan in advance exactly what they're going to buy and remember to bring a range of their own containers with them, while some stores have found that allowing customers to fill their own containers causes spillages and can be unhygienic.

None of this is to say we can't shift to reusables, but when it comes to groceries, prefilled reusable packaging that's professionally washed and refilled by companies is looking like the most promising way forward.

Companies poured millions into shaping our habits and creating new needs. By contrast, their half-hearted attempts so far to spark our interest in reusables that are often less convenient, more expensive and offer little in the way of extra functional benefits have, unsurprisingly, fallen flat. The onus is now on them to devise ways to make reusables more attractive and on regulators to push them to do so in a way that's meaningful.

What is chemical recycling and is it as promising as companies say?

Chemical recycling – called 'advanced recycling' by proponents – breaks down plastic into its chemical building blocks so it can be repeatedly turned into new, high-quality plastic.

A variety of processes fall under this umbrella term. Pyrolysis uses high temperatures to degrade polymers in the absence of oxygen into gas, naphtha, oils and a solid char. Gasification heats plastics in a limited oxygen environment to create 'syngas', which can in turn be converted into chemicals to make new plastics. Solvent-based processes use a solvent to target a polymer, dissolving it so it can be separated to make new plastics.

Proponents of chemical recycling – which include chemical makers and consumer goods companies – say it's the only way to get around the limitations holding back mechanical recycling, which involves cleaning, shredding, melting and re-extruding plastic. The plastic discolours and degrades in quality over time as it's repeatedly recycled. Sometimes it can only be recycled once or twice.

Chemical recycling could also offer a solution for plastic that would otherwise be landfilled or incinerated, like flexible and multilayer plastics. Studies show it generates lower greenhouse gas emissions than using virgin plastic. However, chemical recycling still requires that recyclers get a steady supply of clean, well-sorted material. Most technologies can only handle certain types of plastic, so a big dirty pile of mixed plastic doesn't cut it.

Critics say chemical recycling is little more than smoke and mirrors – it's too energy-intensive and expensive to ever be meaningful, releases toxic emissions, and uses solvents and chemicals that can have negative health impacts. They also say that much of what's touted as chemical recycling is just plastic being turned into fuel rather than new plastic. By one estimate, chemical recycling will only offset 5 per cent of virgin plastic demand by 2040, given the challenges of scaling it.

All this being said, chemical recycling's prospects could improve as the grid decarbonises. Conservation non-profits warn that chemical recycling shouldn't be set up to compete with mechanical recycling by taking higher-quality used plastic – only the real deadbeats that would otherwise go to landfills or incinerators should be chemically recycled. Chemical recycling plants should only be greenlit when it's clear they offer an emissions reduction benefit over virgin plastics and have systems in place to manage any impacts on human health.

The jury is very much still out on chemical recycling. How the chips fall on this one is important. It could decide how reliant on plastics we remain.

Are biodegradable plastics environmentally friendly?

Biodegradable plastics decompose into carbon dioxide, biomass and water. As you know by now, though, the term biodegradable is virtually meaningless when used to market plastics (and, in fact, is banned in some places) since it says nothing about how *long* it takes for something to break down – virtually everything is technically biodegradable if left for long enough.

Compostable plastics, a more defined subset of biodegradable plastics, are plastics designed to break down *under composting conditions in a specific amount of time*.

Some campaigners I've talked to worry that labelling products as biodegradable will encourage people to litter. If littered, biodegradable or compostable plastics will break down far more quickly than conventional plastics, but many people don't realise that the process can still take several years.[2] In the meantime, the packaging can leach plasticisers, stabilisers and other chemicals, entangle animals or be ingested by them.

Today, compostable products and packaging (many marketed as biodegradable), such as cutlery, straws, cups, plates, wipes and dog poop bags, are rarely composted. In the UK, the majority of kerbside-collected food waste goes to anaerobic digestion facilities, which extract energy from the waste but generally reject compostable packaging. In the US, fewer than 5 per cent of the population even has access to kerbside composting facilities for food waste. A 2024 study of US compost facilities showed that 27 per cent said they would accept some compostable food packaging. Not one indicated they'd take non-food compostable packaging.

One reason composters don't want packaging is that regular plastic sneaks in, contaminating entire batches of compost.

---

[2] In a study published in 2019, Richard Thompson and another University of Plymouth researcher exposed compostable bags to air, soil and sea environments similar to ones they'd encounter if littered. They found that compostable bags were still present in the soil after two years. Some bags marketed as biodegradable could still carry shopping three years after being put in soil and seawater.

Compostable non-food packaging also doesn't contribute nutrients to the resulting compost. While some proponents say compostable packaging can help sequester carbon in the soil in the form of biomass, evidence suggests that about half of compostable plastic is lost to carbon dioxide air emissions as it breaks down. Another issue is that compostable packaging that doesn't fully break down could contribute microplastics to the resulting compost.

Brands have dealt with the lack of industrial composting facilities by marketing their products as 'home compostable'. But few people compost in their own gardens; doing so requires a commitment to regularly turning the compost pile to promote aeration, ensuring there's a good balance of nitrogen and carbon and that moisture levels are around 50 per cent. It can take months or even years for compostable plastic to break down at home, and even then, plastic fragments can linger in the compost.

Another tactic brands have used is telling consumers that their biodegradable plastic products will break down in landfills. While lab tests simulating landfill conditions may embolden them to make this claim, it's hard to guarantee that anything breaks down in a real landfill. If it does, it releases methane, a potent greenhouse gas. In the US and many other countries today, some of this methane is captured – but not all. Municipal waste landfills are the third largest source of human-related methane emissions in the US according to the Environmental Protection Agency.

Brands also promise that biodegradable plastics will break down in the oceans, but variations in temperature and micro-organisms in seawater make it very difficult to promise how quickly this will happen. The plastics could hang around for years and, in the meantime, still get stuck in an unsuspecting turtle's nose!

One disposal option in which compostable packaging does have an advantage over conventional is incineration. If incinerated, the net emissions from compostable packaging are likely to be less than for conventional plastic, assuming the compostable material is bio-based.

So are there any applications for which compostable products make sense?

Compostable tea bags, coffee pods, the stickers on fruit peels, ready-meal trays that have food baked on and bin liners for food waste could all make sense. They help ensure more food waste flows to composting facilities, adding nutrients to soil and keeping food out of landfills.

Any product wanting to market itself as compostable should go through an approvals process that requires proof that it does in fact fully break down and that it does so within the timelines industrial composting facilities adhere to. Compostable packaging should only be sold in areas served by industrial composting facilities that say they accept such packaging for composting.

To reduce contamination, one idea I've heard is requiring that all compostable packaging use a unique colour – like bright pink – that no other packaging is allowed to use. Another option is to use compostables in closed venues, like movie theatres, stadiums and airports. In the absence of industrial composting facilities, compostable products marketed as home compostable should be sold only in locations where it is demonstrated that a majority of people are actively composting at home. Compostables should also be labelled with warnings telling consumers that the product won't break down into compost if littered or put in general waste.

Are bio-based plastics a good alternative to fossil-fuel-based ones?

Bio-based plastics are made from biological resources like sugar-cane or corn. They've been around for many decades – cellophane counts as a bio-based plastic – but most haven't been able to compete against fossil-fuel-based plastics on price.

Bio-based plastics aren't necessarily biodegradable. Confusingly, plastics that biodegrade can be made from fossil fuels, and plastics that last for hundreds of years can be made from bio-based materials. About 44 per cent of bioplastics are not biodegradable.

Bio-based plastics can still create many of the same problems as fossil-fuel plastics, and some additional ones. They can take up valuable land or otherwise compete with food. Studies show that some bio-based plastics can take more energy to produce than conventional plastics. Pesticides, PFAs and other chemicals of concern have been found in bioplastics. Some bioplastics – like PLA – can hinder recycling if they mix with PET, the plastic widely used to make drinks bottles.

While bio-based plastics can have far lower carbon emissions than ones derived from fossil fuels, whether they do in practice depends on where they're sourced and how they travel. Locally sourced bioplastics, or ones that travel by sea, will create lower emissions than ones flown in from further away.

Despite the various challenges, incentivising the development of bio-based plastics while seeking to minimise the risks they pose is important. Even PET bottles can only be mechanically recycled a couple of times before they must be downcycled into lower-value products that can't be recycled again, so we're always going to need some virgin plastics.

Scientists are experimenting with making alternatives to conventional plastics from agricultural waste, wood left over from construction projects, seaweed and even carbon dioxide, which can be electrochemically reduced to create oxalic acid that's converted into monomers like glycolic acid to make new plastics. Which materials become commonplace will depend on their relative costs, efficiency and ability to scale. A carbon tax on fossil-fuel-based plastics could help bio-based plastics better compete. Bio-based plastics that offer superior functional benefits to conventional plastics and are easy to recycle could also gain traction.

Bio-based plastics should have a clear end-of-life option before they're launched, whether that be recycling or composting. While we definitely need innovation, the more different materials that companies put on the market, the harder it is to monitor them for chemicals, sort them into separate streams for recycling and find buyers.

Given we burn fossil fuels to generate electricity, isn't making plastics from fossil fuels and then burning the used plastics actually a pretty good idea?

The electricity generated from burning waste is relatively inefficient compared to getting it from other methods. Incinerators are hugely expensive to build and can also create a lock-in effect – they're designed to be fed continuously with waste over a period of about 25 years. That can disincentivise municipalities, who are bound by contracts, from pursuing waste reduction and recycling. Burning used plastics also means that new plastics will need to be made from virgin resources rather than from recycled plastics.

Burning plastic waste releases carbon. It can also lead to the release of dioxins, furans and halogenated flame retardants among other emissions. These are linked to a wide range of health issues, including increasing the risk of heart disease, asthma and cancer. Incinerators also produce hazardous ash that must be safely landfilled. While in wealthy countries, facilities are likely to be closely monitored and emissions cleaned and partly captured, in countries with loose regulations, incinerators could release large amounts of dangerous fumes.

Is landfilling conventional plastics bad for the environment?

For a long time, it's been thought that plastics are inert and so landfilling them doesn't release greenhouse gas emissions or create other harms. That view is starting to change as studies have found microplastics in landfill leachate, and concluded that the additives used to make some plastics can migrate into leachate. If landfills aren't managed properly and if liners erode – which is increasingly likely over time – groundwater can get contaminated.

It's likely that there will always be a need for some landfilling and incineration of plastic waste. In these cases, the focus should be on ensuring this is done safely, that plastics don't escape into the environment and that the leachate and gases from landfills and incinerators are carefully monitored and contained.

Should I worry about chemicals in plastics?

I'm pretty concerned. After talking to scientists in Europe and the US, it's clear that existing regulations are falling well short of protecting us against potentially hazardous chemicals. The fact that plastics can contain multiple chemicals that were never intended to be in there, and for which safety data doesn't exist, is particularly worrying. Even where chemicals are known and regulated, the levels regulators once thought were safe are later deemed to not be safe as new evidence emerges.

A growing body of research shows that exposure to some chemicals like bisphenol A and other bisphenols, even in small doses, can potentially have sizable impacts. While bisphenols get flushed out of our bodies in a few hours, their widespread nature means they're always present in our bodies. Other chemicals, like PFAs, stick around far longer, accumulating in our lungs, liver and other tissues.

Scientists are calling for far stricter regulation, including requiring companies to test mixtures of chemicals that leach from a package rather than simply testing individual chemicals. They also want testing to look for signs of developmental neurotoxicity, immunotoxicity and endocrine disruption, rather than narrowly looking for cancer risks. They want chemicals periodically tested

344

after they're put on the market so that those previously approved can be re-evaluated based on new evidence.

Today – as plastic moves through the supply chain, from resin makers to packaging makers to recyclers and reprocessors – it's often unclear which chemicals have been added at what stage. One idea to address this is introducing a labelling system for packaging – similar to ingredient labels on food – so that packaging makers, recyclers and consumer goods companies know what's in the plastics they're buying.

So far in Europe, reusable packaging must be tested for just three rounds, with the onus being on companies to show that migration of chemicals decreases with each one. Scientists want testing to be expanded to encompass the entire projected time reusable containers will be used for. They say testing should also reflect the conditions under which the products are actually used; for instance, if plastics are put in the dishwasher, they're subject to high heat which can encourage faster degradation.

How can I minimise my exposure to chemicals in plastics?

Since learning more about chemicals in plastics, I buy less canned food and drinks since I know they use plastic liners that can contain bisphenols. I no longer heat food in plastic containers or put plastic in the dishwasher since heat encourages chemicals to leach. I avoid putting hot, greasy and acidic food in plastic for the same reason.

I have replaced my everyday plastic cooking utensils (black ones particularly since they are more likely to be made from e-waste and contain hazardous chemicals) with wooden ones, bought wooden chopping boards and slowly swapped many of my plastic food storage containers for glass ones – though these still use a bit of plastic to seal at the top. I've contacted the suppliers of the non-stick cookware I use to check it doesn't contain PFAs and I use my cast-iron pans more often. I use a stainless-steel reusable water bottle and a ceramic coffee cup rather than plastic ones. I've swapped my kids' plastic cutlery, plates, bottles and cups for stainless-steel ones too. I haven't been able to convince them to stop putting plastic toys in their mouths though, and wooden ones don't last long.

How concerned should I be about microplastics and nanoplastics?

While there's ample evidence that microplastics and nanoplastics harm animal health, their impact on human health is still unclear. There isn't an internationally accepted definition of these plastic particles, while testing methods differ widely, opening researchers up to criticism.

Scientists worry that plastic particles could have far-reaching effects on human health because they act as sponges for chemicals, including those that disrupt the endocrine system. Researchers have also raised concerns that persistent exposure to low levels of microplastics in the air could lead to respiratory and cardiovascular diseases, inflame the gastrointestinal tract and change the intestinal microbiome to create an imbalance between beneficial and harmful bacteria. Beyond the 2024 study I mentioned in Chapter 12 that linked microplastics with an increased risk of humans having a stroke or heart attack, there's ongoing research aimed at uncovering how microplastics impact human health in real life.

Using less plastic food packaging – particularly small packages like sachets, since the ratio of plastic to contents is high – and moving away from plastic kitchenware, like chopping boards, could help. But studies show that microplastics and nanoplastics enter our bodies through a variety of sources, including the soil, air and our skin, so there's not a huge amount any of us can do individually.

Ultimately, the solution lies in regulatory measures that mandate companies to make design changes to products known to leach microplastics – like tyres, paints, synthetic clothes and plastic containers – and put in place filters in washing machines and water treatment plants to stop tiny plastic particles escaping into the environment. As researchers learn more about plastics, they're getting a better sense of the mechanisms that make them degrade. For instance, a 2024 study found that brightly coloured plastics degrade more quickly than white or black ones, indicating that some additives may unintentionally promote degradation.

Are some plastics safer than others?

The science on plastics is still developing, but evidence today suggests that PET, HDPE, LDPE and polypropylene – which go by the resin codes 1, 2, 4 and 5 – are on the safer end of the spectrum, while polyvinyl chloride and polystyrene (which are 3 and 6) have known issues.

PVC can use plasticisers that contain endocrine-disrupting phthalates, while polystyrene has been shown to leach styrene – classified as probably carcinogenic to humans by the World Health Organization – into food and drinks, particularly when used for anything hot or greasy.

The resin code 7 encapsulates a host of other plastics that use various chemicals – including polycarbonates which are known to use bisphenol A – so many public health researchers suggest avoiding plastics with this code.

Many of your answers to these FAQs contain lots of caveats. Is there anything you can say to leave me feeling hopeful?

Yes! For starters, we can rethink the way we do things – it doesn't have to be this way.

We can incentivise companies to be more thoughtful about whether and how they use plastics – and all packaging.

We can use a far smaller subset of materials and use standardised designs that are easier to reuse, recycle and monitor for potential health harms.

We can raise recycling rates by ensuring products are designed for recycling and that companies make new products from recycled content.

We can cut litter – and raise collection rates – by putting a refundable deposit on common packaging.

We can develop better plastics from more environmentally friendly alternatives to fossil fuels.

We can switch away from plastics entirely in some cases where alternatives make sense.

We can develop the infrastructure to handle compostable plastics where these make sense.

We can ensure waste collection has enough funding by getting companies whose products comprise the waste to pay for it.

We can ramp up reusable packaging through the right regulation.

We can use technology to meld the benefits of old business models with the convenience of new ones.

We can shift to packaging that's free of chemicals known to be dangerous and that release fewer microplastics.

We can have systems in place to trap more of the microplastics that are released.

We can continue to be consumers without being consumed.

ACKNOWLEDGEMENTS

Finding a publisher who believed in *Consumed* wasn't easy. In fact, it took a year and a half. During this time, I planned a wedding, had two babies, moved cities, bought a house and kept working for the *Wall Street Journal*, all the while chipping away at my book proposal. The struggle and the wait, at times exhausting, have made this already-precious first book of mine feel even more precious. The biggest thanks go to the people who saw the potential in *Consumed* and took a chance on it.

Thank you to Julia Eagleton who found *Consumed* its publisher, and to Rik Ubhi, formerly of Bonnier, who batted for *Consumed* with an enthusiasm that made me feel re-energised. Thanks to Eric Fletcher, Tim Whiting, Will Francis and Duncan Heath. Thanks to James Lilford, who shepherded this book to completion while being kind and patient throughout.

Thank you to my generous and talented friend Katherine Dunn who offered brilliant suggestions that strengthened *Consumed* and a shoulder to lean on when things felt tough. Thanks also to John Helyar and Denise Roland, both of whom offered advice that made *Consumed* sharper and, mercifully for you, shorter.

I owe a debt of gratitude to the people featured in this book, most of whom submitted to spending hours being interviewed by me and many of whom dug through garages, basements and computers to unearth old documents and photos. This book would not exist without you, and I thank you for letting me record your place in history.

I've turned to countless subject-matter experts over the years to build my knowledge about materials, reuse, recycling, biodegradability, microplastics, chemicals and other topics. I should really thank them all. For this book, however, special thanks goes to Sander Defruyt, Jonathan Scott, Adam Herriott, Dominic Hogg, Morton Barlaz and Ramani Narayan. Thank you also to the Society of Authors whose generous Authors' Foundation grant gave me the extra time I needed to get my book over the line.

Thank you to my parents Rehana and Dhrubo and my brother Rishi, my strongest advocates. To my beautiful children, Rafi and Zara, whose presence underscores every day why our world needs love and protection. To my dog, Suzy, who stayed by my side while I wrote *Consumed* and is the perfect writing companion.

Finally, to my wise and unflappable husband William without whose steadfast support this book would not have seen the light of day. Thank you.

NOTES

Preface

xi **I counted at least eight instances**: Coca-Cola's missed targets include:

1) 1990: Coca-Cola said it would include 25 per cent recycled PET (rPET) in its plastic bottles, a goal it abandoned in 1994.

2) 2000: Coca-Cola said it would have 10 per cent recycled content in 'several billion bottles' that year. Also, in April 2000, Coca-Cola in a consumer advisory said it had been using recycled plastic for the past two years. However, in March 1999, its spokeswoman, Carol Martel, said the company had stopped using recycled plastic in its US bottles in 1994.

3) April 2001: in a memo to bottlers, Coca-Cola said it would have 10 per cent recycled PET in all bottles by 2005. Its 2006 CSR report showed it used just 3.6 per cent rPET in the US.

4) September 2007: Coca-Cola said it would recycle or reuse 10 per cent of its PET bottles in the US but didn't give a date. Neville Isdell, the company's chief executive, told the

Financial Times that the company could achieve its goal within five years. It did not.

5) September 2007: Jeff Seabright told *Plastics News* that Coca-Cola hoped to recycle or reuse 30 per cent of its bottles and cans, and have more than 10 per cent recycled content in its plastic bottles by 2010. The year came and went and Coca-Cola didn't disclose how much recycled content it was using.

6) November 2009: Coca-Cola said it would use 25 per cent rPET by 2015. It didn't meet this target.

7) 2010: Coca-Cola changed its commitment to say it would source 25 per cent of plastic for bottles from recycled or renewable content by 2015. It failed to hit this.

8) 2011: Coca-Cola said it would use plant-based, rather than conventional virgin, plastic in *all* its plastic bottles by 2020. It missed this goal.

9) June 2017: Coca-Cola pledged to increase the recycling rate of its plastic packaging from 14 per cent to 70 per cent by 2025. It is very unlikely to meet this.

xii **That bottled water story:** 'Plastic Water Bottles, Which Enabled a Drinks Boom, Now Threaten a Crisis', Saabira Chaudhuri, *Wall Street Journal*, 12th December, 2018.

xii **Plastic production:** 'New Plastics Economy – Rethinking the Future of Plastics', Ellen MacArthur Foundation, 2016.

xii **Globally, the equivalent:** 'Our Ocean is Choking on Plastic – But It's a Problem We Can Solve', Winnie Lau, Pew Charitable Trust, 8th April, 2022.

xii **Plastic comprises between:** California vs. ExxonMobil, 23rd September, 2024.

xii **the country's recycling rate:** Data is for 2018 from the Environmental Protection Agency.

xii **just 9 per cent:** 'Production, Use, and Fate of All Plastics Ever Made', Roland Geyer et al., *Sci. Adv.3*, e1700782, 2017.

xiii **Sainsbury's announced in 2019:** 'Sainsbury's to halve plastic packaging by 2025', Sainsbury's press release, 13th September, 2019.

xv **the largest per-capita creator:** The US generates 221kg of plastic waste per person according to the OECD's *2022 Global Plastics Outlook*. By contrast, China's plastic waste is 47kg per person and the European Union's is 122kg.

Prologue: The Wandering Garbage Barge

xvii **If he could only:** 'A Mob Boss, a Garbage Boat and Why We Recycle', *Planet Money*, NPR, 10th July, 2019.

xvii **The son of potato and corn farmers:** 'The Reluctant Father of Recycling', Alyson Sheppard, *Lagniappe*, 25th September, 2019.

xvii **He later moved to Houston:** 'It Was Late Last December', Richard C. Firstman, *Newsday*, 11th June, 1987.

xvii **But lately his debts:** Ibid.

xviii **He claimed it was only:** 'A Mob Boss, a Garbage Boat and Why We Recycle', op. cit.

xviii **There he'd found six farmers:** 'Have Barge, Will Travel', Thomas J. Maier, *Newsday*, 8th April, 1987.

xix **'We have enough garbage of our own':** 'Don't Be a Litterbarge', Jacob V. Lamar Jr., *Time*, 4th May, 1987.

xix **The barge lingered:** 'The Garbage No One Wanted', Kerri Jansen, *Waste & Recycling News*, 1st October, 2012.

xix **the *Mobro* was allowed:** 'Trash Odyssey: the Next Move', Shirley E. Perlman, *Newsday*, 17th April, 1987.

xix **We already have deep:** 'It Was Late December', Richard C. Firstman, *Newsday*, 11th June, 1987.

xix **The barge was eventually:** 'Garbage Barge Lost … and Found', Shirley E. Perlman, *Newsday*, 19th April, 1987.

xx **Keeping the boat:** 'Garbage Barge Prods Officials', Philip S. Gutis, *New York Times*, 2nd May, 1987.

xx   **Islip's garbage:** 'L.I.'s Garbage Now an Issue for Diplomats', Philip S. Gutis, *New York Times*, 25th April, 1987.

xx   **Indignant officials:** Ibid.

xx   **The Mexican navy:** 'Mexico Sends Back U.S. Barge Filled With Tonnes of Garbage', *The Gazette*, 27th April, 1987.

xxi   **what movies they watched:** 'Having A Swill Time', Shirley E. Perlman and Phil Mintz, *Newsday*, 5th May, 1987.

xxi   **He longed to return home:** 'Seagoing Garbagemen Say Politics is What Smells', Tom Lowry, *Morning Call*, 7th June, 1987.

xxi   **The captain supported:** 'On Garbage Barge Pride Sustains', Michael Dobbs, *Washington Post*, 9th May, 1987.

xxi   **Harrelson's sister decided:** 'Other News to Note', Bryan Mingle and Clint McCarty, *Orlando Sentinel*, 6th July, 1987.

xxi   **On his 53rd birthday:** 'For Alabamian, L.I.'s Garbage is Dream Gone Bad', Philip S. Gutis, *New York Times*, 6th May, 1987.

xxi   **The rest of the newsroom:** 'Meeting, Memo and Barge Spark Report on Garbage Series', Anthony Marro, *Newsday*, 13th December, 1987.

xxi   **One tenacious reporter:** 'Remember? *Newsday* reporter Shirley Perlman does', Shirley Perlman, *Newsday*, 23rd March, 1997.

xxii   **At sea, eternally:** Song by Jack and Jennifer Sylman, quoted in 'A Few Bars for the Barge', Stuart Vincent, *Newsday*, 12th May, 1987.

xxii   **after 47 days at sea:** 'New York Garbage Barge Bahama Bound?', Associated Press, *Orlando Sentinel*, 7th May, 1987.

xxiii   **I'm losing air space:** 'New York City Shuns Barge Bearing Well-Traveled Trash', Philip S. Gutis, *New York Times*, 16th May, 1987.

xxiv   **a judge ruled:** 'In Search of a Dumping Ground – DAY 142', Peter Marks, *Newsday*, 11th August, 1987.

xxiv   **despite having only:** 'The Rush to Burn, First Part', Irene Virag, *Newsday*, 13th December, 1987.

xxiv **Between 1960 and 1987:** 'The Search for Solutions', Ford Fessenden, *Newsday*, 23rd December, 1987.

xxiv **A study:** 'L.I.'s Garbage Now an Issue for Diplomats', op. cit.

xxiv **The Worldwatch Institute:** 'World Threat of Plastic Trash Defies Technological Solution', Malcolm W. Browne, *New York Times*, 6th September, 1987.

xxiv **wrote Maurice Hinchey:** 'Beyond the Barge: All Our Garbage Will Bury Us', Assemblyman Maurice D. Hinchey, *Newsday*, 13th May, 1987.

xxiv **The very durability:** 'World Threat of Plastic Trash Defies Technological Solution', op. cit.

xxiv **America, the richest nation:** 'What Did We Learn From the Garbage Barge?', Frank R. Jones, *Newsday*, 13th August, 1987.

Chapter 1: Throwaway Living

3 **Stouffer told the room:** 'Plastics Packaging: Today and Tomorrow', Lloyd Stouffer, National Plastics Conference, November 1963.

4 **in the 25 years:** *Better Living* magazine, November–December 1962.

6 **Starting in 1915:** *American Plastic: A Cultural History*, Jeffrey L. Meikle, Rutgers University Press, 1997.

7 **Girl Collapses:** *DuPont Magazine*, September–October 1944.

7 **DuPont hired writers:** *DuPont: From the Banks of the Brandywine to Miracles of Science*, Adrian Kinnane, Johns Hopkins University Press, 2002.

7 **By 1940:** *DuPont Magazine*, 140th anniversary issue, 1942.

8 **The raw materials:** Ibid.

9 **We are a strong nation:** Ibid.

9 **the plastics industry's sales:** 'WPB Reports Output of Chemicals Increased 90% in War Years', *Wall Street Journal*, 24th September, 1945.

9   **His answer:** *Life in Plastic: It's Fantastic*, Robert Edwards and Rachel Kellett, Other India Press, 2000.

10   **a new tube of toothpaste:** 'Remembering American Wars From the Home Front', Beth Dippel, *Sheboygan Press*, 12th November, 2016.

10   **Utilities companies:** 'Home Repair Shops', Harry T. Rohs, *Wall Street Journal*, 1st March, 1943.

10   **Make it do:** *DuPont Magazine*, April–May 1944.

10   **Envelopes were reused:** 'On the Business Horizon: Packages in War Dress', *Barron's*, 26th January, 1942.

11   **DuPont housewives:** *Better Living*, July–August 1957.

11   **the largest influx of women:** 'The Rise and Fall of Female Labor Force Participation During World War II in the United States', Evan K. Rose, *The Journal of Economic History*, September 2018.

12   **dwarfing any previous:** *Better Living*, January–February 1956.

12   **the number of married women:** 'New Twists in Packaging', David A. Loehwing, *Barron's*, 22nd June, 1959.

12   **It buoyantly described:** *Better Living*, May–June 1960.

13   **Some customers considered:** *Better Living*, July 1947.

13   **Over the next two decades:** Ibid.

13   **DuPont's data showed:** *DuPont Magazine*, April 1947.

13   **A new atmosphere:** *DuPont Magazine*, October 1941.

13   **The men asked housewives:** *Better Living*, July–August 1954.

14   **half-slices of bacon:** Ibid.

14   **portions of coffee:** *DuPont Magazine*, March 1946.

14   **There were early signs:** *DuPont Magazine*, May 1941.

14   **Cleanliness and Cellophane:** *DuPont Magazine*, August–September 1946.

15   **Wrapping produce cost more:** 'Packaging: Transparent Wrapper's Popularity Soars', Lee Geist, *Wall Street Journal*, 26th March, 1954.

16   **It's just a matter of time:** Ibid.

16 **It cost just 2 cents:** Ibid.

16 **90 per cent of fresh carrots:** Ibid.

16 **A Los Angeles supermarket chain:** 'Some Shoppers Howl as More Fresh Foods Show Up in Packages', Harold D. Watkins, *Wall Street Journal*, 5th January, 1959.

16 **Meat packed in polyethylene:** 'Plastic Film: It Stars in Packaging Parade', Felicia Anthenelli, *Wall Street Journal*, 25th April, 1953.

16 **the freezer was the fastest-growing:** *Better Living*, September–October 1952.

16 **demand soared:** 'Chicken Dinners', Winston C. Fournier, *Wall Street Journal*, 17th February, 1954.

17 **Clarence Birdseye:** *Plastic: The Making of a Synthetic Century*, Stephen Fenichell, HarperCollins, 1996.

17 **They offered lower prices:** *DuPont Magazine*, November–December 1948.

17 **Whether management likes it or not:** 'Coffee in Offices', Caroline Bird Menuez, *Personnel Journal*, May 1950.

18 **The drink began appearing:** 'Bus Man's Battle', Mitchell Gordon, *Wall Street Journal*, 17th March, 1950.

18 **its volume exceeding:** 'Lily-Tulip Cup Net Put At $7.25 A Share', *Barron's*, 12th November, 1951.

18 **Americans were using:** 'Sip and Flip: Low-Cost Plastic Cups Challenge Disposable Paper Type', *Wall Street Journal*, 29th February, 1956.

18 **In offices and canteens:** 'Container War: Plastic Food Packaging Makes Deeper Inroads on Paper', Bernard Wysocki Jr, *Wall Street Journal*, 8th May, 1978.

19 **allowing manufacturers to slash costs:** Ibid.

19 **Cookies and chocolates:** 'Our Daily Bread: Nutrition Experts Say Americans Are Eating Themselves to Death', Mary Bralove, *Wall Street Journal*, 6th January, 1971.

19 **98 per cent of Americans:** Ibid.

20 **America's teenage population:** *Better Living*, November–December 1963.

20 **a whopping 7 per cent:** 'Skeleton at the Feast?', John Slatter, *Barron's*, 3rd October, 1966.

20 **raw plastic resin:** 'Breaking the Mold', Dana L. Thomas, *Barron's*, 26th March, 1962.

20 **up from just 130 million pounds:** 'Shaping the Future', Lawrence Armour, *Barron's*, 6th June, 1960.

20 **enthused the vice president:** Ibid.

20 **Polyethylene's price dropped:** *Better Living*, November–December 1962.

22 **the US was generating:** *Better Living*, March–April 1970.

22 **DuPont's vice president:** *Better Living*, July–August 1968.

22 **By the decade's end:** *American Plastic: A Cultural History*, op. cit.

22 **Attendees were asked:** Proceedings: First National Conference on Packaging Wastes, San Francisco, 1969.

24 **Commoner headed:** 'City Studies Little Dioxin From Garbage-Fueled Plants', Matthew L. Wald, *New York Times*, 26th September, 1984.

24 **At hearings that followed:** 'With No Room at the Dump, U.S. Faces a Garbage Crisis', Philip Shabecoff, *New York Times*, 29th June, 1987.

24 **Incineration represents the ultimate:** Proceedings: First National Conference on Packaging Wastes, op. cit.

Chapter 2: The Million Dollar Boys' Club

26 **Carson had quipped:** 'Heeeere's Johnny's Solution', Joshua Quittner, *Newsday*, 26th April, 1987.

27 **It was humorous:** Author interview with Steven Englebright.

28 **any other type of packaging:** Minutes from Environment and Energy Committee Meeting of the Suffolk County Legislature

on 31st August, 1987 at the William H. Rogers Legislature Building, Hauppauge, New York.

28 **The proposal, if approved:** 'Putting Lid on Plastic Waste: Suffolk bill would mandate biodegradable packaging', Bill Bleyer, *Newsday*, 14th August, 1987.

29 **His brief was clear:** Author interview with Donald Shea.

29 **factories were working full tilt:** 'Plastic Producers' Stocks Are Outperforming the Market and Signal Continuing Success', Barry Meier, *Wall Street Journal*, 25th March, 1987.

31 **the plastics industry's annual revenue:** 'Opposition to Plastic Packaging is Intensifying as the Nation's Solid-Waste Problem Grows Acute', Elliott D. Lee, *Wall Street Journal*, Eastern edition, 25th November, 1987.

32 **In Shea's first year:** Interview with Don Shea.

33 **compared to 29 per cent:** 'Plastic, Plastic and More Plastic', Susan Youngwood, *Vermont Business Magazine*, November 1988.

34 **Given its members:** Author interview with Ron Liesemer. I spoke with Ron in 2021 and again in 2022. He passed away on 4th June, 2023, at the age of 85, a few days before we were scheduled to speak again.

35 **If we didn't get in quickly:** Author interview with Ron Liesemer.

35 **Back in 1985:** 'The Environment', Leo H. Carney, *New York Times*, 15th September, 1985.

35 **Coke at the time:** 'Coca-Cola Begins Testing Plastic Cans', *The Gazette*, 15th October, 1985.

36 **They sold plastics recycling machines:** 'Plastic Recycling Turns Soft-Drink Bottles Into – Ski Parkas?', Peter Tonge, *Christian Science Monitor*, 28th July, 1987.

36 **The industry must stop:** 'Plastics Recycling Grows Up', Nicholas Basta, *Chemical Engineering*, 23rd November, 1987.

37 **In the words of Henry Ford:** 'Cash From Mountain of Trash Recyclers Panning for Profits in Castoff Plastics', Anne Murphy, *Newsday*, 24th July, 1989.

37 **predicting that plastics:** 'Drops in the (Plastic) Bucket?', Richard Saltus, *Boston Globe*, 27th August, 1990.

37 **says Dennis McKeever:** Author interview with Dennis McKeever.

38 **More recycling programmes:** Ibid.

38 **A memo to the industry:** 'Plastics Deteriorate in Image Only', Jack Anderson and Dale Van Atta, *Washington Post*, 30th January, 1990.

Chapter 3: McToxic

40 **They rolled out:** *Fast Food Nation*, Eric Schlosser, Houghton Mifflin, 2001.

40 **Imagine – No Carhops:** Ibid.

41 **Labour costs plummeted:** Ibid.

41 **the paper didn't retain:** 'Battle of the Clamshell', Jeb Blount, *Globe and Mail*, 15th March, 1991.

41 **Studies showed that:** 'New Ecological Symbol to Promote Recycling', Container Corporation of America, 6th September, 1970.

41 **In large part to allay:** 'Pollution Prevention in Corporate Strategy, Case B1: The Clamshell Controversy', National Pollution Prevention Center for Higher Education, March 1995.

41 **The polystyrene clamshell:** *Barefoot to Billionaire*, Jon Huntsman, Abrams, 2014.

41 **Staff could quickly:** *The Battle to Do Good*, Bob Langert, Emerald Publishing, 2019.

41 **When the company went:** Author interview with Shelby Yastrow.

42 **It's estimated that:** 'The Rise and Rise of Big Mac', Steven Greenhouse, *New York Times*, 8th June, 1986.

42 **120 million pounds of lettuce:** 'Kellogg Keeps New Cereals Coming', George Lazarus, *Chicago Tribune*, 21st February, 1989.

43 **Globally, a new McDonald's restaurant:** 'Fast-Food Leader: McDonald's Combines a Dead Man's Advice With Lively Strategy', Robert Johnson, *Wall Street Journal*, 18th December, 1987.

43 **Made from the chemical styrene:** 'PS, I Love You: Polystyrene is Winning Friends in Packaging, Furniture and Other New Markets', Neil A. Martin, *Barron's*, 14th October, 1968.

43 **900 million coffee cups:** 'Cleaning Up: Reducing, reusing or recycling waste saves money and preserves the environment', Kimberley Noble, *Globe and Mail*, 6th February, 1989.

43 **By the time the *Mobro* set sail:** 'Tarnish on the Golden Arches', Karen Stults, *Business and Society Review*, Fall 1987.

43 **breakfast items, drinks and salads:** Interview with Shelby Yastrow.

43 **We're going to be wrapped up:** 'The Search for Solutions: State and Federal Government are Focusing Their Attention on Garbage and Considering How to Cut Waste as Well as Burn and Bury it', Ford Fessenden, *Newsday*, 23rd December, 1987.

44 **He took the job:** 'From Fast Food to Fast Cars – Retired McDonald's CEO Revs Up Career in Racing', Anna Marie Kukec, *Daily Herald*, 25th August, 2000.

44 **Colleagues soon nicknamed him:** '"McHistory" Observed at Corporate Gala', Mark Potts, *Washington Post*, 21st November, 1984.

45 **If you've got time to lean:** 'You'll Find It's Mean to Lean at McDonald's', Janet Key, *Chicago Tribune*, 10th October, 1989.

45 **He was fiercely loyal:** 'From Fast Food to Fast Cars – Retired McDonald's CEO Revs Up Career in Racing', op. cit.

45 **including pushing people:** 'Our Mission, The Boardroom Initiative, https://boardroominitiative.org/our-mission/

45 **The partner introduced:** Author interview with Shelby Yastrow.

47 **The McDonald's president's orders:** Rensi's reluctance to let go of the container was described in an interview by Bob Langert and confirmed in an interview with Yastrow.

47 **Alinsky introduced me:** Author interview with William Collette.

48 **the dingy surroundings:** Author interview with William Collette.

49 **likely transmitted by hamburger fat and drinks:** Interview with, and witness statement from, Brian Lipsett, who worked at the Citizens Clearinghouse as a research analyst through the McToxics campaign.

50 **it was only making the change:** 'McDonald's Tries to Help Save Ozone', Associated Press, *Orlando Sentinel*, 6th August, 1987.

51 **the fifth largest amount of toxic waste:** Witness statement from Brian Lipsett.

51 **McDonald's began flying Rathje:** 'McDonald's: The Greening of the Golden Arches', William Gifford, *Rolling Stone*, 22nd August, 1991.

51 **Rathje, through his digs:** *Rubbish! The Archaeology of Garbage*, William Rathje and Cullen Murphy, HarperCollins, 1992.

51 **four billion of which:** 'The Greening of Corporate America', *The Record*, 27th January, 1991.

51 **aerate the soil:** 'Taking Pollution Out of the Bag', Peter M. Gianotti, *Newsday*, 23rd August, 1989.

52 **In April of that year:** 'McDonald's Speaks Out', *Restaurant Business*, 20th May, 1989.

53 **One survey showed:** 'The Rise and Rise of Big Mac', op. cit.

53 **About a third:** 'Fast-Food Leader: McDonald's Combines a Dead Man's Advice With Lively Strategy', op. cit.

54 **Switching to paper:** 'Wrapped Up in Debate on Plastic, Suffolk bill is touted and trashed', Thomas J. Maier, *Newsday*, 10th April, 1988.

54 **McDonald's was already grappling:** 'McDonald's Sees Nothing to Beef About – It's Full Speed Ahead for the Fast-Food Giant', Richard Gibson, *Wall Street Journal*, 1st April, 1988.

54 **We can put the McDLT in paper:** 'McDonald's Speaks Out', op. cit.

54 **The teacher encouraged:** 'Pupil Starts a Revolution of Her Own: Ban Plastics', George James, *New York Times*, 28th April, 1989.

56 **Think about that:** 'Fighting McLitter', Mike Kelly, *The Record*, 27th April, 1989.

56 **I know how easy:** 'Students Find Adults Dislike McPlastics', Pat R. Gilbert, *The Record*, 11th June, 1989.

56 **We're not going to put:** 'Kids Wage a Big Mac Attack on Packaging', Pat R. Gilbert, *The Record*, 12th May, 1989.

56 **while just 82 people:** 'Students Find Adults Dislike McPlastics', op. cit.

57 **What you may not know:** 'Kids' Order: Hold the Plastic', Tom Toolen, *The Record*, 9th February, 1990.

57 **Legislatures in 26 US states:** 'Huntsman Fighting Foes of Plastic Packaging', Joel Campbell, *Deseret News*, 30th November, 1988.

57 **told a foodservice magazine:** 'Environmental Economics Will Dictate Operational Plans', Monica Kass, *Restaurants & Institutions*, 27th November, 1989.

59 **I vowed then:** *The Battle to Do Good*, op. cit.

59 **There was no way:** Interview with Patrick Halpin, the Suffolk County executive at the time.

59 **The company tracked:** *The Battle to Do Good*, op. cit.

60 **But the resulting quality:** 'A Setback for Polystyrene', John Holusha, *New York Times*, 18th November, 1990.

61 **all the chain's restaurants:** 'McDonald's Aims to Recycle Trash, Image', Judith Crown, *Crain's Chicago Business*, 16th April, 1990.

61 **the fast-food chain's environmental activities:** 'Captive Kids: A Report on Commercial Pressures on Kids at Schools', *Consumer Reports*, 1st January, 1998.

61 **defended its use:** 'Corporate Presence in Schools Grows', Randall Lane, *Chicago Tribune*, 18th November, 1990.

62 **but offered nothing else:** 'Get Advertising Out of Our Schools', Charlotte Baecher, *USA Today*, 20th July, 1990.

62 **Early brand loyalty:** Ibid.

62 **Greenpeace described greenwashing:** 'The Earth's Own Day: Shall We Celebrate or Deride Our Efforts to End the Torment of Our Planet? Greenwashing', Andrew Davis, *Newsday*, 15th April, 1990.

62 **Ed Rensi:** 'World Wildlife Fund, McDonald's Announce Largest Environmental Education Effort Ever for Nation's School System', PR Newswire, 9th March, 1990.

63 **It was conceived:** 'McDonald's Oak Brook Architect Reflects on Proposed Move Downtown in Changing Times', Samantha Bomkamp, *Chicago Tribune*, 8th June, 2016.

64 **He told Yastrow:** The account of this meeting is from an interview I conducted with Bob Langert.

64 **It cut packaging:** 'McDonald's Speaks Out', op. cit.

65 **You aren't going to turn:** As told to me by Bob Langert.

65 **About 70 per cent of fast-food purchases:** *Fast Food Nation*, op. cit.

65 **The non-profit had avoided:** Interview with Fred Krupp.

66 **We did it for the good earth:** *ABC World News Tonight*, 1st November, 1990.

67 **The wrap's quilt-like effect:** 'McDonald's to Drop Plastic Foam Boxes in Favor of High-Tech Paper Packaging', Scott Kilman, *Wall Street Journal*, 2nd November, 1990.

67 **We wonder what the planet:** 'Burger King's Ad Gives Backhanded Salute to Rival', *Wall Street Journal*, 8th November, 1990.

67 **Making one A4 sheet:** 'Disposable Paper-Based Food Packaging – The False Solution to the Packaging Waste Crisis', Manon Stravens, *Profundo*, 28th July, 2023.

67 **Paper production uses:** 'Overview of Intentionally Used Food Contact Chemicals and Their Hazards', Ksenia J. Groh et al., *Environment International*, May 2021.

67 **In time, the paper food packaging:** 'Dangerous PFAS Chemicals Are in Your Food Packaging', Kevin Loria, *Consumer Reports*, 24th March, 2022.

68 **Researchers have raised concerns:** 'Study: "Safer" PFAS in Food Packaging Still Hazardous', O'Neill School, Indiana University, 29th March, 2023.

68 **Although McDonald's was careful:** 'McDonald's to Drop Plastic Foam Boxes in Favor of High-Tech Paper Packaging', op. cit.

68 **Leftover food is even more damaging:** Interviews with Jonathan Scott, DS Smith's Kemsley mill technical operations director, and Timothee Duret, director of sustainable technology at Veolia UK.

68 **Under pressure from non-profits:** 'McDonalds's Promises to Eliminate Foam Packaging by 2019', *As You Sow*, 10th January, 2018.

69 **McDonald's has scored an environmental touchdown:** 'McDonald's to Phase Out Foam Packaging', press release, 1st November, 1990.

69 **The chain alone:** 'Topics of The Times: Greening of the Golden Arch', *New York Times*, 7th November, 1990.

69 **Others followed suit:** 'Walking on Eggshells, Polystyrene People Make a Comeback', Jeff Bailey, *Wall Street Journal*, 21st January, 1999.

69 **It's a big blast:** 'McD's Switch to Paper Fuels Packaging Debate', Alan Liddle, *Nation's Restaurant News*, 19th November, 1990.

Chapter 4: Talking Shit

71 **Some environmentalists estimated:** 'What's New in Diapers; Cloth Fights for Favor as Disposables Fill the Dumps', Barnaby J. Feder, *New York Times*, 12th March, 1989.

76 **Environmentalists used Lehrburger's study:** 'Diaper Wars!', Jeanne Wirka, *Environmental Action*, March/April 1989.

76 **Today it is emerging:** 'Consumer's World: Do Disposable Diapers Ever Go Away?', Michael deCourcy Hinds, *New York Times*, 10th December, 1988.

77 **As a society:** 'Diapers in the Waste Stream', Carl Lehrburger, January 1989.

77 **The cloth diaper rental:** 'Dy-Dee Wash Looks Ahead', *Kiplinger Magazine: the Changing Times*, Vol. 1, Iss. 3, March 1947.

77 **By the time the Second World War ended:** Ibid.

78 **Getting diapers washed commercially:** Ibid.

78 **Mills was apparently shocked:** 'Pampers: The Birth of P&G's First $10 Billion Dollar Brand', P&G press release, 27th June, 2012.

78 **My mother says:** Author interview with Gracie Schild.

78 **no doubt inspired by his grandchildren:** 'Victor Mills, a "Pampered" Career', Claudia Flavell-While, *The Chemical Engineer*, 19th January, 2018.

78 **Disposable diapers had in fact existed:** 'The Bottomline on Disposables', Lawrence E. Joseph, *New York Times*, 23rd September, 1990.

79 **But they were expensive:** 'Pampers: The Birth of P&G's First 10-Billion-Dollar Brand', P&G, 27th June 2012, https://www.pg.co.uk/blogs/pampers-birth-pgs-first-10-billion-dollar-brand/

79 **The researchers delighted:** 'Smaller', Malcolm Gladwell, *The New Yorker*, 26th November, 2001.

79 **Instead, the researchers designed:** 'Disposable Diaper' patent, R.C. Duncan Etal, 27th April 1965, https://patentimages.storage.googleapis.com/c6/ca/b3/029007572b981f/US3180335.pdf

79 **Mills sent his team back:** 'Consumer Choice, The Driving Force of Market Economy', P&G booklet; 'Pampers: A Case Study'.

80 **Test that diaper:** *Eyes on Tomorrow: The Evolution of Procter & Gamble*, Oscar Schisgall, Doubleday, 1981.

80 **Overwhelmingly, mothers said:** 'Consumer Choice, The Driving Force of Market Economy. Pampers: A Case Study', op. cit.

80 **One engineer described it:** *Eyes on Tomorrow*, op. cit.

81 **At 10 cents apiece:** 'Pampers: The Disposable Diaper War', Mark Parry and M.L. Martin, Darden Case No. UVA-M-0643, 20th June, 2001.

81 **Spending on new equipment:** 'Procter & Gamble Outlays Cut Rise in 1st Half Net', *Wall Street Journal*, 15th February, 1962.

81 **Just a few years after:** *Eyes on Tomorrow*, op. cit.

82 **Scott Paper killed the diaper:** 'Marketing Classic: How P&G Put the Squeeze on Scott Paper Co.', Jim Hyatt and John E. Cooney, *Wall Street Journal*, 20th October, 1971.

83 **Nice to know:** 'The Warming Climate', Jim Wechlser, *Brattleboro Reformer*, 11th June, 1970.

83　**America's largest advertiser:** 'P&G Boosts Nontraditional Marketing: Move Reflects Effort to Target Specific Groups', Alecia Swasy, *Wall Street Journal*, 25th November, 1988.

83　**In 1970 alone:** 'Disposable Diaper War Loses a Competitor, Namely Scott Paper', *Wall Street Journal*, 30th June, 1971.

83　**By the early 1970s:** 'Time for a Change? Unseat Disposables, Diaper Services Cry', John L. Moore, *Wall Street Journal*, 17th August, 1978.

83　**Pampers had become:** 'Seven Decades of Disposable Diapers', Davis Dyer, EDANA, August 2005.

83　**The company had:** 'Disposable Diaper War Loses a Competitor, Namely Scott Paper', op. cit.

84　**It was also eyeing:** 'Procter & Gamble Chief Doesn't Expect Price Controls to Impede Earnings Growth', *Wall Street Journal*, 14th May, 1973.

84　**He added stand-up elasticised flaps:** Opinion in Kimberly Clark vs. Procter & Gamble, 26th August, 1992.

84　**Basically, the message:** Author interview with Wayne Sanders.

85　**The popularity of the new Huggies:** 'P&G's New Disposable Diaper Intensifies Marketing Battle With Kimberly-Clark', John Bussey, *Wall Street Journal*, 4th January, 1985.

85　**P&G was very, very conservative:** Author interview with Terry Batsch.

85　**Huggies grabbed market share:** 'Personality Change For P&G', Winston Williams, *New York Times*, 23rd March, 1984.

85　**Smith also poached:** 'Paper Tiger: How Kimberly-Clark Wraps Its Bottom Line in Disposable Huggies From Newsprint and Tissues', Cynthia F. Mitchell, *Wall Street Journal*, 23rd July, 1987.

85　**Finding pregnant women:** 'Baby-Goods Firms See Direct Mail as the Perfect Pitch for New Moms', Bob Davis, *Wall Street Journal*, 29th January, 1986.

85 **It included not just their names:** 'Newspapers Fear Being Bypassed by Advertisers', Thomas B. Rosenstiel, *Los Angeles Times*, 27th April, 1989.

85 **Kimberly-Clark also bought:** 'Baby-Goods Firms See Direct Mail as the Perfect Pitch for New Moms', op. cit.

86 **By the end of 1985:** 'P&G Pins Down Lead in Sales of Disposable Diapers', *Chicago Tribune*, 14th July, 1986.

86 **diapers were P&G's most important product:** 'Personality Change For P&G', op. cit.

87 **Once wetted:** 'Description of gel blocking etc. from judge's opinion in Civ. A. No. 2:87-0047-1', Procter Gamble Co. vs. Kimberly-Clark, 740 F. Supp. 1177, D.S.C. 1989.

87 **It absorbed up to 80 times:** 'Superabsorbent Diapers: Marketers Seek Doctors' Support Amid Health Concerns', Jolie Solomon, *Wall Street Journal*, 5th September, 1986.

88 **Wall Street analysts:** Ibid.

88 **These diapers are the biggest success:** 'New Weapons in Diaper War', Steven Greenhouse, *New York Times*, 25th November, 1986.

88 **Supermarkets got in on the act:** 'P&G's New Disposable Diaper Intensifies Marketing Battle With Kimberly-Clark', op. cit.

88 **By 1987:** 'Consumer's World; Do Disposable Diapers Ever Go Away?', op. cit.

89 **Before the end of the decade:** 'Diaper Wars!', op. cit.

89 **In all likelihood:** 'Consumer's World; Do Disposable Diapers Ever Go Away?', op. cit.

89 **Procter & Gamble spends more money:** 'Disposable Diapers Near Saturation Point?', Herb Greenberg, *Chicago Tribune*, 7th August, 1984.

89 **In 1988:** 'Makers of Adult Diapers Foresee Expanding Market as Population Ages', Mark McLaughlin, *New England Business*, 1st February, 1988.

90 **Fast forward a few years:** 'China Sees Two Sides of Open-Crotch Pants for Babies', Associated Press, 26th November, 2003.

91 **There was a lot of environmental advocacy:** Author interview with Tim Sergeant.

92 **The storied institution:** 'Washington Talk: The National Press Club', Barbara Gamarekian, *New York Times*, 24th March, 1988.

93 **If P&G can find a way:** 'Recycle Diapers? Good – If P&G Does It', *Newsday*, 27th June, 1989.

93 **The Boston-based company:** 'Battle of the Grocery Bags: Plastics Versus Paper', Lisa Belkin, *New York Times*, 17th November, 1984.

93 **That study hit:** 'Vinyl-Chloride Risk Was Known by a Lot of People Even Before the First Deaths', Barry Kramer, *Wall Street Journal*, 2nd October, 1974.

93 **Using Arthur D. Little's assumptions:** 'How "Tactical Research" Muddied Diaper Debate', Cynthia Crossen, *Wall Street Journal*, 17th May, 1994.

94 **Unfortunately, many comments on this issue:** 'Update on Diapers', Center for Policy Alternatives, September 1990.

95 **Although P&G had contributed:** 'Research Funded By P&G Attacks Cloth-Diaper Use', Michael Waldholz, *Wall Street Journal*, 10th April, 1991.

95 **The first-ever LCA:** 'LCA – How it Came About', Robert G. Hunt and William E. Franklin, *The International Journal of Life Cycle Assessment*, March 1996.

95 **Comparing different variables:** 'How "Tactical Research" Muddied Diaper Debate', op. cit.

96 **Lawmakers also proposed:** 'P&G Gets Mixed Marks as It Promotes Green Image but Tries to Shield Brands', Alecia Swasy, *Wall Street Journal*, 26th August, 1991.

96 **Proposals in California:** 'States Mull Rash of Diaper Regulations', Kathleen Deveny, *Wall Street Journal*, 15th June, 1990.

97 **The company sent:** Ibid.

97 **I have never seen an issue:** 'Leaders of Plastics Industry Seeking 25% Recycling Goal', John Holusha, *New York Times*, 29th March, 1991.

98 **In 1989 alone:** 'Can Business Save Our Schools?', Eleanor Branch, *Black Enterprise*, March 1991.

98 **It also omitted:** 'P&G's Lessons on Environment Provoke Criticism', David E. Kalish, *Marketing News*, 14th March, 1994.

98 **Even members of the advisory panel:** 'Environmentalists Question Value of Procter Gamble Teaching Kit', David E. Kalsih, *Journal Record*, 22nd January, 1994.

98 **Over the next two years:** Ibid.

99 **The paper fibres were sucked out:** 'Process May Clean Up on Dirty Diapers – Project is Producing Paper Fiber', Dick Lilly, *Seattle Times*, 27th July, 1990.

99 **Relying on the seats:** 'Recycling Project Truly In Its Infancy – Disposable-Diaper Experiment Begins' Constantine Angelos, Seattle Times, 23rd February, 1990.

99 **While disposable diapers:** 'Diaper-Recycling Looks Like Technological Success, Economic Flop', Constantine Angelos, *Seattle Times*, 25th January, 1991.

99 **It was nowhere near enough:** 'The Bottomline on Disposables', op. cit.

100 **It appears that the costs:** 'Diaper-Recycling Looks Like Technological Success, Economic Flop', op. cit.

100 **Norma McDonald was sitting:** 'Artzt Tells P&G Shareholders Company Taking Lead on Key Environmental Solutions', PR Newswire, 9th October, 1990.

100 **At that time, boy:** Author interview with Norma McDonald.

101 **These are carbon-carbon backbone polymers:** Author interview with Chuck Pettigrew.

101 **A P&G executive even wrote:** 'USCC: 25 Years and Growing' USCC, https://www.compostingcouncil.org/page/silveranniversary

102 **But, by 1972:** 'Reclaiming Rubbish', Burt Schorr, *Wall Street Journal*, 2nd June, 1972.

102 **Compost couldn't compete:** Ibid.

102 **To make it economically viable:** 'Diapers Get Smaller, Not Compostable', BioCycle, *Emmaus*, Vol. 43, Iss. 1., January 2002.

102 **We called them the 'goofies':** Author interview with Jim McNelly.

103 **P&G began running TV ads:** 'P&G to Tout Disposable Diaper Benefits', Dick Rawe, *Chicago Tribune*, 21st July, 1991.

104 **While we encourage companies:** 'Procter & Gamble Campaign Halted in Green-Collar Fraud Settlement', *News from Attorney General Robert Abrams*, 14th November, 1991.

104 **P&G had also handed out:** 'Crazy for Cloth: The Benefits of Cotton Diapers', *Contemporary Women's Issues Database*, 1st January, 2003.

105 **By 1996:** 'Marketing: P&G Fetes 35 Years Without Diaper Pails', *Dayton Daily News*, 8th September, 1996.

106 **that means just:** 'Two Experts Do Battle Over Potty Training', Erica Goode, *New York Times*, 12th January, 1999.

Chapter 5: Hefty's Claims Turn to Trash

111 **Degradable was a feel-good word:** Author interview. Mike Levy passed away in February 2024. My conversations with him for this book took place over phone and email through the second half of 2023.

113 **Although Mobil had for years:** 'Back to the Lab', Amal Kumar Naj, *Wall Street Journal*, 21st July, 1988.

114　**New equipment required:** Ibid.

114　**The chemical companies' lobbyists:** 'Dilemma for Makers of Plastics', John Donnelly, Associated Press, 6th November, 1989.

114　**There is no such thing:** 'Is Degradability a Throwaway Claim?', Barry Meier, *New York Times*, 17th February, 1990.

115　**we don't want to be behind that:** 'Answers Sought to Mounting Plastic', Elizabeth A. Brown, *Christian Science Monitor*, 11th July, 1989.

115　**And it's that public relations value:** 'The Green Revolution: Mobil', Jennifer Lawrence, *Advertising Age*, 29th January, 1991.

115　**But by the 1960s:** 'Garbage, by the Numbers', *Chicago Tribune*, 10th October, 2003.

115　**US sanitation departments:** *American Plastic: A Cultural History*, Jeffrey L. Meikle, Rutgers University Press, 1997.

116　**Pack up your troubles:** '1978 Glad 2-Ply Trash Bags "Pack up your troubles" TV Commercial', YouTube, https://www.youtube.com/watch?v=d81XJAs1pK8

116　**Between 1960 and 1990:** Municipal Solid Waste Management data from the Environmental Protection Agency.

116　**Never touch garbage again!:** 'December 1987 – WNYW Commercial Break for "It's Howdy Doody Time"', YouTube, https://www.youtube.com/watch?v=KvxFj9-1HgM

117　**Ruffies' sales jumped:** 'Modern-day Alchemy Transforms Plastic to Gold', Dick Youngblood, *Minneapolis Star and Tribune*, 4th August, 1986.

117　**Poly-Tech promised:** 'Degradable Bags are Touted as Way to Take out Trash', Pete Leffler, *Morning Call*, 24th October, 1988.

118　**He also underscored:** 'Ruffies to Offer Degradable Plastic Trash Bags Next Month', Ingrid Sundstrom, *Star Tribune*, 14th October, 1988.

118 **The money is much better spent:** 'Degradable Bags Sold Nationally', *Indiana Gazette*, 19th November, 1988.

118 **A small diaper maker:** 'Corn Growers Hope to Play Major Role in Degradable Plastics', Sharon Schmickle, *Star Tribune*, 12th April, 1989.

118 **Mobil's marketers floated the idea:** Interview with Mike Levy.

119 **By 1990:** 'Mobil to Recycle Plastic Grocery Bags', *Globe and Mail*, 17th May, 1990.

119 **But plastic bags:** 'Raging Battle: Paper Bag vs. the Plastic Bag', Terence Roth, *Wall Street Journal*, 25th February, 1985.

120 **We're talking out of both sides:** 'Biodegradable Trash Bag Just Marketing Tool', *The Gazette*, 24th September, 1989.

121 **We were proud:** Author interview with Hubert H. Humphrey II.

121 **Between 1989 and 1991:** 'Evaluation of Environmental Marketing Terms in the United States', EPA, February, 1993.

121 **A September 1990 survey:** 'Working with Environmental Groups', James T. Harris, *Public Relations Journal*, 1st May, 1992.

122 **He asked if:** 'Mobil Says Earnings Fell 3.2%', Thomas C. Hayes, *New York Times*, 24th January, 1990.

122 **In the decade to 1987:** 'FTC Reasserts Itself, Wins Headlines for New Activism', Mark Potts, *Star Tribune*, 30th June, 1991.

122 **there was a mass exodus:** 'FTC Regulatory Activity and the Information Content of Advertising', Avery M. Abernethy et al., *Journal of Public Policy & Marketing*, Fall 1998.

123 **It's not fat-free:** 'Texas Law Official Stirs Up Marketers', Christi Harlan, *Wall Street Journal*, 6th June, 1991.

124 **Once at the FTC:** 'Remarks of Janet D. Steiger Chairman, Federal Trade Commission, Before the American Advertising Federation Government Relations Conference', 12th March, 1991.

127 **Ross Love:** Janet Steiger speech, EPA National Environmental Labelling Conference, 1st October, 1991.

127 **Another ad exec:** Ibid.

127 **Our pitch to industry:** Author interview with Bob Viney.

127 **After months of work:** 'Industry Seeks U.S. Rules Covering Environment Ads', John Holusha, *New York Times*, 15th February, 1991.

128 **Everyone will know it:** Tom Rattray speech at EPA Environmental Labelling Conference, 2nd October, 1991.

128 **Mary Engle:** Author interview with Mary Engle.

128 **When consumers were subsequently polled:** COPE WAVE Survey, March 1993, as referenced in the letter from the attorneys general to the FTC, 28th September, 1995.

128 **Adding a qualifier:** Ibid.

129 **The new national certainty:** Environmental Defense Fund comments to FTC, 27th November, 1995.

130 **As long as the products:** Recommendations from State AGs to the FTC, *Green Guides Review*, Matter No. P954501, 24th April, 2023.

Chapter 6: Plastics Make it Possible

131 **Each time, plastics were in place:** 'DMB&B Makes Plastic Personal', Lucas Sloane, *Adweek*, 11th January, 1999.

132 **Sixty-four per cent of Americans:** *APC: The Value of Membership*, Publication of the American Plastics Council, 1995.

133 **The public's approval:** Interview with Jean Statler.

133 **The rest of you don't:** This account was described by Don Olsen, who was present at the Greenbrier meeting.

133 **too poor to afford indoor plumbing:** *Deseret News* dialogue with Jon Huntsman Sr., Paul S. Edwards, 30th October, 2011.

134 **Some people called him:** 'Plastics Downturn Whets a Big Appetite', Richard Koenig, *Wall Street Journal*, 6th November, 1989.

134 **When McDonald's switched:** Wirthlin Worldwide Application for ARF David Ogilvy Research Award, American Plastics Council Case, October 1996.

134 **Once those three do it:** Interview with Ron Yocum.

135 **Trained as an economist and statistician:** 'Richard Wirthlin, Pollster Who Advised Reagan, Dies at 80', Adam Clymper, *New York Times*, 17th March, 2011.

135 **Sitting next to Reed:** *The Greatest Communicator*, Dick Wirthlin, Wiley & Sons, 2004.

136 **a fish swimming in a bowl of wet cement:** Ibid.

136 **Reagan also asked:** Ibid.

136 **He almost gave me an ulcer:** Ibid.

137 **Wirthlin summed up:** Interview with Jean Statler, who thought this was Wirthlin's mantra when in fact it was Reagan's.

137 **the prince of pollsters:** C-SPAN, 7th February, 1992.

137 **pollster general:** 'The Pernicious Power of the Polls: How pollsters and marketing gurus are measuring your choices – and shaping them', Brock Brower, *CNN Money*, 1st March, 1988.

137 **By 1992:** 'Powerful New Group Will Take Over from CSWS', Robert D. Leaversuch, Modern Plastics, December 1991.

137 **but also P&G:** P&G's continued involvement was mentioned by P&G's John Pepper, Quantum's Ron Yocum and APC's Don Shea.

138 **The ad also eventually:** 'Industry Settles Complaints on Plastics Ads', Scott Allen, *Boston Globe*, 21st December, 1995.

139 **Wirthlin agreed on two main goals:** Wirthlin Worldwide Application for ARF David Ogilvy Research Award American Plastics Council Case, op. cit.

139 **In July 1992:** 'APC Takes Laurels for Ad Campaign Research', Steve Toloken, *Plastics News*, 28th April, 1997.

139 **Understanding the negatives:** Wirthlin Worldwide Application for ARF David Ogilvy Research Award American Plastics Council Case, op. cit.

141 **When the public learned:** Ibid.

141 **Simultaneously, though:** 'Environment', Frank Edward Allen, *Wall Street Journal*, 18th December, 1992.

142 **Quantum's milk bottle recycling plant:** 'Quantum Considers Exiting Recycling', *Plastics News*, 22nd May, 1995.

142 **They had a vibrating feed system:** Author interview with George Rizzo.

142 **By 1993:** 'Plastics: Conventional Recycling Moves Downstream', Andrew Wood and Elizabeth S. Kiesche, *Chemical Week*, Vol. 153, Iss. 7, 25th August/1st September, 1993.

142 **They described how plastic:** 'Environment', op. cit.

143 *Plastics News* **reported:** 'Rutgers' Recycling Center Gets the Ax', Roger King, *Plastics News*, 2nd September, 1996.

Chapter 7: The Chasing Arrows

144 **By 1987:** Author interview with Lewis Freeman.

146 **Recycling can save:** *Examining Material Evidence: The Carbon Fingerprint*, Imperial College, 2020.

148 **Sponsored by Container Corp:** 'New Ecological Symbol to Promote Recycling', Container Corporation of America, 16th September, 1970.

148 **I was kind of an arrogant little punk:** Author interview with Gary Anderson.

148 **In announcing the new design:** Ibid.

149 **It was a marketing tool:** Ibid.

150 **That's number 6:** 'Our Cup Runneth Over: Detergent Measuring Cups are Another Wasteful Pollutant', Nancy Durnford, *The Gazette*, 4th November, 1989.

150 **A Heinz spokesman:** 'Plastics Industry Uses Marketing Tactic to Tell Public Products are Recyclable', Bill Paul, *Wall Street Journal*, 28th July, 1989.

150 **Container coding:** 'Mauro Lauds Plan to Make Recycling of Plastics Easier', Jim Phillips, *Austin American-Statesman*, 6th January, 1989.

151 **Companies eagerly embraced the law:** 'Eight Recyclers, Enviros Seek to Ban Chasing Arrows From Packaging', Steven Lang, *Integrated Waste Management*, 20th July, 1994.

151 **Surveys showed:** WAVE V survey as referenced in Council on Packaging in the Environment comments to the FTC, 9th October, 1995. The Council in its comments raised concerns about the validity of the questions included in the survey, saying they were confusing and subject to acquiescence bias.

152 **The consumer was blaming us:** Author interview with Coy Smith.

152 **One afternoon, Bradley answered the phone:** Author interview with Athena Bradley.

153 **Machinery would need to be changed:** 'Summary Report: SPI Resin Identification Code Final Recommendations', 2nd May, 1993.

154 **Both men thought:** Author interviews with Terry Guerin and Mark Lichtenstein.

155 **SPI's president Larry Thomas:** Letter from Mark Lichtenstein to Keith Atkins of Union Carbide, 25th January, 1995.

156 **Even their rates are poor:** 'New Association of Plastic Recyclers State-of-the-Industry Report Shows Strength of US Plastic Recycling', APR Press Release, 9th August, 2022.

156 **The overall recycling rate:** Data as of 2018 from the Environmental Protection Agency.

156 **Nearly 70 per cent:** 'Reduce. Reuse. Confuse.', Consumer Brands Association, 2019.

Chapter 8: The Garbologist

158 **Rathje counted Procter & Gamble:** 'William L. Rathje, Father of Garbology', Arizona Anthropologist Centennial, Michael Brian Schiffer, 2015.

158 **Bill was fairly conservative:** Author interview with Michael Brian Schiffer.

158 **Dig a trench:** 'Rubbish!', William L. Rathje, *Atlantic Monthly*, December 1989.

158 **While food and yard waste did degrade slowly:** Dan Grossman and Seth Shulman, 'Down in the Dumps', *Discover*, April 1990, 36–41.

159 **It was a big hit:** Author interview with Don Olsen.

161 **He often greeted perplexed students:** 'William Rathje (July 1, 194 – May 24, 2012)' Studio Michael Shanks – Stanford, 7th June, 2012, https://web.stanford.edu/group/archaeolog/cgi-bin/archaeolog/2012/06/07/william-rathje-july-1-1945-may-24-2012/

161 **A reporter for the *Boston Globe*:** 'Reading our Garbage', Mark Muro, *Boston Globe*, 10th August, 1981.

161 **Given that there was no digging involved:** 'A Conversation with William Rathje', Matthew R. Lane, *Anthropology Now*, April 2011.

163 **He once tried to set up a garbage museum:** 'Archaeologist Digs in Dumps, Buries Environmental Theory', Miles Corwin, *Los Angeles Times* via the *Akron Beacon Journal*, 18th July, 1993.

163 **In later years:** 'He Loves Trash', Nick Casey, *Stanford Magazine*, November/December 2006.

164 **What we're trying to do:** 'Garbage and American Society' C-Span, 27th August, 1990, https://www.c-span.org/video/?13700-1/garbage-american-society

164 **It was true that:** 'Municipal Solid Waste in the United States, 2007 Facts and Figures', US Environmental Protection Agency.

164 **One theory was:** 'Waste of a Sort: Curbside Recycling Comforts the Soul, But Benefits Are Scant', Jeff Bailey, *Wall Street Journal*, 19th January, 1995.

164 **Between 1984 and 1994:** 'Advancing Sustainable Materials Management: 2018 Fact Sheet', US Environmental Protection Agency.

164 **I am not worried:** 'Rubbish!', *Atlantic Monthly*, op. cit.

165 **His landfill digs:** 'Mining Trash Heap for Answers to Riddles', Richard Severo, *New York Times*, 5th October, 1989.

165 **Because the Garbage Project:** 'William L. Rathje, Father of Garbology', op. cit.

165 **In a 1993 newspaper interview:** 'The Rotten Truth About Garbage', Miles Corwin, *Los Angeles Times*, 17th July, 1993.

165 **He's out advocating things:** 'Telltale Trash', Rex Springston, *Richmond Times-Dispatch*, 30th April, 1998.

166 **In addition to espousing:** Rathje, in his book, cites Consumer Reports and the Office of Technology Assessment to make this claim; however, a study published in the *Journal of Pediatrics* in 1979 found the opposite.

167 **So if you've got kids at home:** 'Diapering the Dumps. Disposables Soak Up Toxic Waste, Scientist Says', Kimberly Noble, *Globe and Mail*, Toronto, 30th May, 1990.

167 **Let's deal with the big-ticket items:** 'Among the Earth Baby Set, Disposable Diapers Are Back', Michael Specter, *New York Times*, 23rd October, 1992.

168 **And if we don't think:** 'Among the Earth Baby Set, Disposable Diapers Are Back', op. cit.

168 **The economics of fast food recycling:** Suffolk County hearing, August 1987.

169 **Foam trays cost:** 'Wrapped Up in Debate on Plastic, Suffolk bill is touted and trashed', Thomas J. Maier, *Newsday*, 10th April, 1988.

169 **At a hearing:** 'Halpin Hears Plastic Ban Debate', Rick Brand, *Newsday*, 12th April, 1988.

169 **Kevin Dietly:** Suffolk County hearing, 23rd February, 1993.

169 **an establishment could comply:** 'Gaffney Approves New Plastics Ban', Katti Gray, *Newsday*, 6th April, 1994.

170 **The word 'recycling':** 'Cafeterias Adapt to Recycling Law', Regina Marcazzo, *New York Times*, 17th July, 1994.

170 **Rathje's research showed that:** 'Our Landfills Runneth Over', Erin Ellis, *Edmonton Journal*, 21st February, 1993.

170 **His shtick:** Author interview with Robert Lilienfeld.

171 **The utility of legislated source reduction:** 'Rubbish!', *Atlantic Monthly*, op. cit.

172 **The seminars reached:** 'Flexible Packaging Association Offers Area Teachers a Fresh Approach to Science, Environmental Studies', *PR Newswire*, 4th April, 1997.

172 **The value of plastics:** 'King o' the Landfill Hill: Rathje Talks Trash, As Told to Roger Renstrom', *Plastics News*, 8th March, 1999.

172 **Today, flexible and multilayer plastics:** 'Breaking the Plastic Wave: A Comprehensive Assessment of Pathways Towards Stopping Ocean Plastic Pollution', PEW Charitable Trusts and SystemIQ, 2020. (To qualify, this is 80 per cent of macroplastics.)

173 **Around 25,000 flexible packages:** 'The Global Commitment Five Years On', Ellen MacArthur Foundation, October 2023.

173 **plastic waste per person:** 'The Real Truth About the U.S. Plastics Recycling Rate', Beyond Plastics and The Last Beach Cleanup, May 2022.

Chapter 9: Making a Market of the World's Poorest

177 **In 1990:** 'Plastic Units Plead for Long-Term Policy', *Times of India*, 2nd February, 1990.

178 **People who had long brought their own bags:** 'Say "No" to the Plastic Bag, Urge Greens', Sharmila Joshi, *Times of India*, 4th June, 1995.

178 **By 1997:** 'Watch Out, It's Plastic, Stupid', Bhavana Padivath, *Times of India*, 22nd March, 1998.

179 **The worry, even if it was there, was quite low:** Author interview with Vivek Bali.

180 **They cut bars of soap:** 'The Man Behind the Sachet Revolution in India', R.V. Rajan, *Madras Musings*, Vol. XXX, No. 17, 1st–15th January, 2021.

180 **There was a realisation:** Author interview with Vindi Banga.

181 **It is not necessarily the rich woman:** Speech of the Chairman at the Annual General Meeting, Hindustan Lever Ltd, 8th April, 1958.

181 **All consumer goods manufacturers:** Ibid.

181 **You need to convince consumers:** Remarks by Peter ter Kulve at Unilever investor seminar, December 2014.

182 **They performed skits:** 'Hindustan Lever Ltd. Vice-Chairman's Speech', *Times of India*, 19th April, 1961.

182 **Operating and maintaining the vans:** 'Rural Marketing in India. Speech of the Chairman Shri V.G. Rajadhyaksha', *Times of India*, 25th April, 1969.

182 **The cost of marketing:** Ibid.

182 **The efforts didn't justify.** Ibid.

183 **You can buy one banana:** Author interview with Anand Kripalu.

184 **India's loose-leaf tea sellers:** Author interview with Shiv Shivakumar.

184 **He later shifted:** Author interview with C.K. Ranganathan, son of Chinni Krishnan and founder of CavinKare.

185 **We applied shampoo:** Ibid.

188 **A 1980s advertisement:** Author interview with Alec Lever.

189 **By 1989:** 'When Villages Awake', *Times of India*, 25th August, 1990.

190 **Suddenly the population woke up:** Author interview with Sanjiv Kakkar.

190 **China and Indonesia's per-capita shampoo consumption:** 'Giants Battle for Foam & Froth Market', Ritwika Chaudhuri, *Economic Times*, 15th May, 1997.

192 **They overspent their ad budget:** Author interview with Udaiyan Jatar.

193 **It was steadily gaining ground:** 'Lever to Launch Lux Range of Shampoos', *Business Standard*, 14th May, 1997.

194 **In 1996 alone:** 'Grooming for Growth', Chayya, *Business Standard*, 7th September, 1997.

194 **Between 1993 and 1996:** 'Elephants Can't Dance', S.L. Rao, *Economic Times*, 26th July, 1999.

194 **By the late 1990s:** 'Strategic Innovation: Hindustan Lever Ltd', Rekha Balu, *Fast Company*, 31st May, 2001.

194 **India accounted for:** 'Indian Villagers Turn on to Brand Names', Anatharaman Muralikumar, Reuters, 8th September, 1999.

195 **This was particularly the case:** 'Marketers Need to Participate in Rural Growth', Business Line (*The Hindu*), 6th October, 2001.

195 **It promised to give Indians:** 'HLL Unveils Gold Promo for Shampoo Sachets', *Times of India*, 7th September, 2001.

196 **One senior executive:** Author interview with Shiv Shivakumar.

196 **By 2002, shampoo penetration:** 'Brand-Equity – Sachet Up the Ramp', *Economic Times*, 13th March, 2002.

197 **India's rural population:** 'Strategic Innovation: Hindustan Lever', Rekha Balu, *Fast Company*, Boston, Iss. 47, June 2001.

197 **It arranged loans:** 'Hindustan Lever's Project Shakti – Marketing FMCG to the Rural Consumer', V. Kasturi Rangan, Harvard Business School, 23rd March, 2006.

198 **He made the case:** *The Fortune at the Bottom of the Pyramid*, C.K. Prahalad, FT/Prentice Hall, 2004.

199 **Selling shampoo in rural India:** 'The Pyramid and the Pie: Corporate Involvement in Sustainable Development', Atul Wad, Sustainablebusiness.com, December 2004.

200 **the *Lucknow Times* published:** 'Cows Fall Prey to Polythene Bags', Manjari Mishra, *Times of India*, 1st April, 2000.

201 **India imported 19 million kilograms:** 'India as a Toxic Dump', Praful Bidwai, *Times of India*, 9th October, 1995.

201 **Every so often:** 'Taken to Task – India Forms a Task Force to Manage Plastic Wastes', *Asian Chemical News*, 8th November, 1996.

201 **Knowing the importance of the cow:** Author interview with M.K. Sharma.

202 **Nihal Kaviratne:** Author interview with Nihal Kaviratne.

203 **sachets accounted for 90 per cent:** 'HLL's Shampoo Offers to Alter Sachet-Bottle Ratio to 60:40', *Financial Express*, 17th March, 2004.

203 **Chik's owner:** 'FMCG Cos Change Distribution Tack to Woo Consumers', Sindhu J. Bhattacharya, Business Line (*The Hindu*), 9th March, 2005.

204 **The sachets attracted lower sales taxes:** 'Sachets: The Next Big (Small) Thing', Ravi Balakrishnan and Aarti Razdan, *Economic Times*, 29th June, 2005.

205 **99 per cent were sachets:** 'Small Formats and Sachets', Confederation of Indian Industry, January 2024.

Chapter 10: Farewell to the Age of Ecology
206 **Within hours of each other:** 'Cola Companies Plan to Use Recycled Plastic in Bottles', *Vancouver Sun*, 5th December, 1990.
207 **We needed to do something:** Author interview with Deborah Cross.
207 **In March 1991:** 'Recycled Plastic Bottles Tested', *The Gazette*, 13th March, 1991.
207 **At the time:** 'U.S. Plastics Firms Plan 25% Recycling', *Kitchener–Waterloo Record*, 30th March, 1991.
207 **It said it would keep rolling:** 'Coke & Pepsi Go Flat on Recycled Content', Sarah S. Smith, *Plastics News*, 8th March, 1999.
207 **Within a couple of years:** 'Dear Coke: Start Recycling', Nancy Vogel, *Sacramento Bee*, 14th November, 1998. From GRRN Scrapbook.
207 **oil prices had dropped:** Ibid.
210 **The ten US states with bottle bills:** 'Vermont Senator Packages Bottle Bill', Danielle Jackson, *Waste Age*, Overland Park, Vol. 33, Iss. 6, June 2002.
210 **Taxpayers and local governments:** Letter from Lance King to Coca-Cola, 19th March, 1997.
210 **Bottle bills are taxation:** Author interview with Doug Ivester.
210 **Container deposit laws:** Coca-Cola 10-K for year ended 31st December, 2008.
211 **The deposits were as high:** *Citizen Coke*, Bartow J. Elmore, W.W. Norton, 2015.
211 **Then, in 1935:** 'Competitive Position of Owens-Illinois Glass', Craig MacDonald, *Barron's*, 30th November, 1942.

211 **Thick returnable glass bottles:** 'Tin Can Invasion of Soft Drink Field Speeds Up', Kenneth S. Smith, *Wall Street Journal*, 20th July, 1961.

211 **Its executives believed:** Ibid.

211 **They also observed:** 'Canned Soda Pop', T.A. Wise, *Wall Street Journal*, 10th May, 1954

211 **Prohibition had seen:** 'The American Beverage Industry and the Development of Curbside Recycling Programs 1950–2000', Bartow J. Elmore, *The Business History Review*, Autumn 2012.

211 **Canned beer grew:** 'The Coming Battle of the Packages', *Nation's Business*, Vol. 32, Iss. 11, November 1944.

212 **Americans watching popular TV shows:** 'Tin Can Invasion Of Soft Drink Field Speeds Up', op. cit.

212 **Coke eventually started:** *Citizen Coke*, op. cit.

213 **Lured by the convertibles:** *The Birth of a Salesman: Ernest Dichter and the Objects of Desire*, Daniel Horowitz, self-published, 1986.

213 **Americans were now buying:** 'Preliminary Final Report on the Introduction of the No-Return Bottle in the Soft Drink Industry', Ernest Dichter, June 1961.

214 **The new packaging:** 'A Motivational Study on the Introduction of the No-Return Bottle to Soft Drink Bottlers, Retailers and Consumers', Ernest Dichter, November 1961.

214 **The steady drop:** 'More Throw-Aways; Glass Container Makers Have Barely Scratched the Surface of the Market', Norris Willatt, *Barron's*, 22nd August, 1966.

214 **They had long fumed:** 'Tin Can Invasion of Soft Drink Field Speeds Up', op. cit.

214 **By 1966:** 'More Throw-Aways', op. cit.

214 **By the time Lance King:** *Citizen Coke*, op. cit.

215 **Between 1955 and 1969:** 'The Beverage Container Problem', Environmental Protection Agency, September 1972.

215 **By the early 1970s:** 'The Battle Over Bottles & Cans', *Changing Times*, December 1971.

215 **More than 350 bills:** 'The Beverage Container Problem', op. cit.

216 **the year before the ad first ran:** 'Monsanto Co. Expects Fourth Quarter Net To Fall "20% or More"', *Wall Street Journal*, 22nd December, 1969.

216 **They quickly caught on:** 'Plastic Bottle: It Wins Converts from Glass as Output Soars', T.A. Wise, *Wall Street Journal*, 11th September, 1950.

216 **plastic bottles were cheap enough:** 'Shaping the Future: Plastic Fabricators are Turning Out a Growing Variety of Products', Lawrence Armour, *Barron's*, 6th June, 1960.

216 **Sohio sent 5,000 letters:** 'Sohio Vexed by Moves to Ban No-Returns, Pushes Plastic Bottle', Jim Hyatt, *Wall Street Journal*, 4th March, 1971.

216 **He also claimed that:** 'Self-Destruct Bottle May Replace Can Ban', *Akron Beacon Journal*, 16th February, 1971.

216 **Sohio even told journalists:** 'Ban Won't Cut Litter', *Akron Beacon Journal*, 9th March, 1971.

216 **A new plastic bottle:** 'Pepsi to Test New Disposable Plastic Bottle', *Daily Reporter*, 4th April, 1970.

217 **Market research showed:** 'Sohio Vexed by Moves to Ban No-Returns, Pushes Plastic Bottle', op. cit.

217 **The bottle was 17 ounces lighter:** 'Coca-Cola Unveils Its Plastic Bottle, Made by Monsanto', *Wall Street Journal*, 4th June, 1975.

217 **Japanese workers exposed to it:** 'Ban on Plastic Soft Drink and Beer Bottles Sought in Suit by Environmental Group', *Wall Street Journal*, 22nd April, 1976.

217 **NRDC maintained:** Ibid.

218 **Returnable glass made up less:** 'Damn, This is Different', Larry Jabbonsky, *Beverage World*, 3rd June, 1995.

218 **The man told Ivester:** 'Built to Last', Ed Morales, Terry College of Business, University of Georgia, 8th November, 2019.

219 **I need you back here:** Author interview with Doug Ivester.

219 **While the company was doing brisk business overseas:** 'Coca-Cola Continues Strong Gains as Sales Surge in Foreign Markets', Reuters and other reports, *Toronto Star*, 18th July, 1990.

219 **McWhorter had gone:** Author interview with Susan McWhorter (now Susan Driscoll).

220 **Younger consumers saw the bottle:** 'Coca-Cola Introduces Marketing Gimmick: Its Famous Old Bottle', Michael J. McCarthy, *Wall Street Journal*, 12th January, 1993.

220 **We were really training consumers:** Author interview with Susan McWhorter.

221 **Coke will loan you the money:** As recounted by Susan McWhorter in an interview with the author.

221 **Sales jumped 25 per cent:** 'Coke Hypes Contour in Post Promo', Karen Benezra, *Brandweek*, 11th April, 1994.

221 **The new 20-ounce package:** 'A Change of Face', *Sun-Sentinel* wire services, 12th January, 1993.

221 **The plastic contour bottle:** Coca-Cola annual report, 1993.

221 **The *Wall Street Journal*:** 'Coca-Cola Introduces Marketing Gimmick: Its Famous Old Bottle', op. cit.

222 **We call it unleashing:** 'Damn, This is Different', op. cit.

222 **The results were beyond anything:** Coca-Cola annual report, 1994.

222 **By September 1994:** 'Best Sales Gain in 5 Years Likely at Coca-Cola', Cynthia Mitchell, *Atlanta Constitution*, 20th September, 1994.

222 **Coca-Cola forecast that total sales:** 'A Worth Vessel', John N. Maclean, *Chicago Tribune*, 25th May, 1994.

222 **Contour is a brand:** 'Behemoth on a Tear', Maria Mallory, *Bloomberg Businessweek*, 3rd October, 1994.

222 **Through the 1990s:** Beverage Marketing Corp data.

222 **Coca-Cola was selling an estimated 20 million sodas each day:** 'Growing National Campaign Challenges Coca-Cola Chairman Ivester to Cut Plastic Soda Bottle Waste', *PR Newswire*, 12th November, 1998.

222 **The recycling rate for plastic bottles:** Data provided by Container Recycling Institute.

223 **Targeting Coke had helped GRRN:** 'Grants Give Grassroots Crusade More Muscle', Steve Toloken, *Plastics News*, 3rd August, 1998.

225 **We did not break a pledge:** 'Newspaper Letters, Faxes & E-Mail: Coke Didn't Break Vow', Jeff Foote, *Atlanta Constitution*, 22nd December, 1999.

225 **The company told concerned consumers:** 'Coke's Broken Promise', Dave Aftandilian, *Conscious Choice*, February 2000.

225 **But non-profits estimated:** 'Coke's Broken Promise', op. cit.

225 **Our position is that:** 'Activists Urge Coke to Recycle Plastic', Valerie Santillanes, *Albuquerque Journal*, 16th September, 1998.

226 **It was the first time:** Author interview with Steve Navedo.

227 **What the hell are you doing?:** Author interview with Pierre Ferrari.

230 **The rate was half:** 'Report Shows Plastic Bottle Waste Tripled Since 1995', Container Recycling Institute, 15th September, 2003.

Chapter 11: Taken Captive

232 **It emerged that:** 'Olympian Drama: A Preliminary Report on Salt Lake Scandal Certain to Rile the IOC', Craig Copetas and Roger Thurow, *Wall Street Journal*, 20th January, 1999.

232 **News outlets reported:** 'Olympic Question: When is a Gift a Bribe?' Larry Siddons, Associated Press, 24th December, 1998.

233 '**The company spent $300,000:** Winter Olympics 2002 Recycling Update', Gary Liss, GRRN, 21st January, 2002.

234 **Every night:** 'Planners Outline Salt Lake City's Olympic Budget', Associated Press, 1st May, 2001.

234 **After months of looking:** Author interview with Kate Krebs.

236 **Families of loggers:** Author interview with Ed Boisson.

237 **I think it's fair to say:** Author interview with Mark Loughmiller.

239 **As we got closer:** Author interview with Julie Seitz, known back then as Julie Wilsterman.

240 **Other NRC board members:** As per interview with Meg Morris, who at the time served as NRC's board president.

240 **The companies behind BPEC:** The MOU talks about increasing recycling 'through innovative comprehensive' programmes, which is industry speak for kerbside recycling that takes all materials, rather than container deposit laws which single out drinks containers.

240 **Many of them belonged:** The National Soft Drink Association changed its name to the American Beverage Association in 2004 to reflect the much wider range of drinks its members now sold, such as bottled water, fruit juices, iced teas and sports drinks.

240 **I expect great things:** 'Mailbag', Kate Krebs, *Plastics News*, 28th June, 2004.

242 **Drinks bottles were widely littered:** 'Diapers in the Waste Stream', Carl Lehrburger, January 1989.

243 **In 2001, bottled water sales grew:** 'In a Water Fight, Coke and Pepsi Try Opposite Tacks', Betsy McKay, *Wall Street Journal*, 19th April, 2002.

243 **Haight worked for:** NYPIRG was one of the largest of the Ralph Nader-inspired public interest research groups that began

cropping up across college campuses in the 1970s. Students elected board members, pooled their money, picked issues they cared about and hired lobbyists to fight their corner.

245 **We realise that home:** 'Beverage Council's Report Negates Single-Serve Theory', Joe Truini, *Waste News*, 19th September, 2005.

245 **The report, for which NRC paid:** Form 990 filed by the National Recycling Coalition for 2005.

245 **The Container Recycling Institute:** Container Research Institute Winter newsletter, 2005.

246 **In addition to the questions:** Ibid.

246 **Do we really want:** 'Can't Contain Her Frustration', Pat Franklin, *Waste & Recycling News*, 5th June, 2006.

246 **Kate Krebs herself:** 'Beverage Council's Report Negates Single-Serve Theory', op. cit.

246 **NRC's credibility is utmost:** 'Experts Lament State of Container Recycling', Steve Toloken, *Plastics News*, 16th May, 2005.

248 **It really made people mad:** Author interview with Jerry Powell.

249 **Krebs was being compensated:** Form 990, National Recycling Coalition for fiscal year ending 31st March, 2007.

249 **She migrated towards:** Author interview with George Dreckmann.

250 **The rent was $163,254:** NRC report and financial statements draft, 31st March, 2008 and 2007.

250 **In 2007, Coke announced:** 'Coca-Cola Sets Goal to Recycle or Reuse 100 Percent of Its Plastic Bottles in the U.S.', *Business Wire*, 5th September, 2007.

251 **A newspaper article about the Coke plant:** 'Coca-Cola to Help Build Plastic Recycling Plant', Jeff Nesmith, Cox Washington Bureau, *Atlanta Journal-Constitution*, 6th September, 2007.

251 **the latest available data:** 'Coke's "Green" Effort Includes More Cash for RecycleBank', Theresa Howard, *Daily Times*, 6th September, 2007.

251 **In April 2008:** 'Coca-Cola's Green Crusader', Marc Gunther, *Fortune*, 28th April, 2008.

251 **The bottle was made of glass:** Data from Beverage Marketing Corp.

252 **Clearly it was in their best interests:** Author interview with Ed Skernolis.

252 **Particularly as public opinion:** Written Testimony of Joseph K. Doss President and CEO International Bottled Water Association Before the Subcommittee on Transportation Safety, Infrastructure Security, and Water Quality of the Environment and Public Works Committee United States Senate Hearing on Quality and Environmental Impacts of Bottled Water, 10th September, 2008.

253 **This would prevent:** 'Coca-Cola, URRC Open World's Largest Plastic Bottle-to-Bottle Recycling Plant', Coca-Cola press release, 14th January, 2009.

254 **Notably, Coca-Cola's plan:** 'Coke Set to Open JV PET Recycling Plant', Mike Verespej, *Plastics News*, 16th January, 2009.

256 **Reporters who quoted Krebs:** 'Coca-Cola Set to Start Up Recycling Plant on Jan. 14', Trevor Anderson, *Herald-Journal*, 6th January, 2009.

256 **The Climate Group:** 'American Beverage Association and The Climate Group Announce Powerful Partnership', *Bevnet*, 19th December, 2008.

256 **To see Coca-Cola:** 'Coca-Cola Goes Green With Bottle Recycling Plant', Trevor Anderson, *Herald-Journal*, 18th January, 2009.

256 **Coca-Cola's media relations department:** '*PR News* Announces CSR Award Finalists', *PR News*, 12th January, 2009.

257 **That single initiative:** 'Creating Value for Business and Society', The Brookings Institution, 13th May, 2009.

257 **Samples were readily available:** 'Big Red Goes Completely Green at Olympics', Natalie Zmuda, *Advertising Age*, 1st February, 2010.

257 **Coke vowed that:** 'Coca-Cola Sees Green', Jennifer Zegler, *Beverage Industry*, June 2011.

257 **Scott Vitters went a step further:** 'The Coca-Cola Company's New PlantBottle® Packaging On Store Shelves Nationwide Today!', *Businesswire*, 4th April, 2011.

257 **Coke's research showed that:** 'Coca-Cola Opens Recycling Plant in Spartanburg', Trevor Anderson, McClatchy, *Tribune Business News*, 15th January, 2009.

258 **The company said its aim:** 'New Plant to Help Coke Meet Recycling Goals', Mike Verespej, *Plastics News*, 19th January, 2009.

258 **We took this hideous material:** Author interview with Jeff Seabright.

258 **You melt crap:** Author interview with Carlos Gutierrez.

259 **The plant was burning:** The figure was provided by a former Coca-Cola executive who was directly involved and requested anonymity because he isn't permitted to disclose the information.

259 **More than half of those:** 'China to Allow Whole PET Scrap Bottle Imports', Steve Toloken, *Plastics News*, 14th December, 2009.

259 **Coke brought in another company:** Author interview with Mitch Hecht, who ran the International Recycling Group. 'We were losing too much money too rapidly,' he said.

260 **The percentage of PET bottles:** Data provided by NAPCOR. The data shows PET bottles sold in the US and recycled food-grade PET in the US and Canada.

260 **Fourteen months after Krebs left:** 'A Very Quick Exit; NRC's Abrupt Closing Spurs Hopes for a New Beginning', Bruce Geiselman, *Waste & Recycling News*, 14th September, 2009.

260 **Its debt stood:** 'NRC's Debt Outlook Brightens', Amanda Smith-Teutsch, *Waste & Recycling News*, 2nd August, 2010.

260 **One described the organisation:** 'Members Ponder NRC's Fate', Joe Truini, *Waste News*, 13th April, 2009.

260 **The cancellation of the congress:** 'NRC: One Last Shot? Past Leaders Make 11th-hour Bid to Stave Off Coalition's Demise', Jim Johnson, *Waste & Recycling News*, 28th September, 2009.

260 **Coke took its money:** 'KAB Takes Over Coca-Cola Partnership', Amanda Smith-Teutsch, *Waste & Recycling News*, 15th February, 2010.

260 **Container deposits lead to higher costs:** 'Indiana Lawmakers Consider Recycling Law', Nathan Laliberte, *American Metal Market*, 12th April, 2013.

261 **Today, more than five decades after:** 2022 data from NAPCOR.

Chapter 12: The Last Straw

266 **You can find it online:** 'Sea Turtle with Straw up its Nostril – "NO" TO SINGLE-USE PLASTIC' YouTube, https://www.youtube.com/watch?v=4wH878t78bww

268 **Over the next five years:** 'The Business Case for a UN Treaty on Plastic Pollution', WWF, the Ellen MacArthur Foundation and BCG, October 2020.

269 **A whopping 79 per cent:** 'Production, Use, and Fate of All Plastics Ever Made', Roland Geyer et al., *Science Advances*, 19th July, 2017.

269 **Consumer goods companies:** 'The Business Case for a UN Treaty on Plastic Pollution', op. cit.

269 **The presence of small plastic particles:** 'Plastics on the Sargasso Sea Surface', Edward J. Carpenter et al., *Science*, 17th March, 1972, and 'Plastic Particles in Surface Waters of the Northwestern Atlantic', J.B. Colton Jr. et al., *Science*, 9th August, 1974.

269 **I first found it:** 'A Place to Chill Out Between summers', Glenn Gaslin, *Los Angeles Times*, 7th March, 2002.

269 **In it he described:** 'Lost at Sea: Where Is All the Plastic?', Richard C. Thompson et. al., *Science*, Vol. 304, Iss. 5672, 7th May, 2004.

269 **Thompson wrote a second paper:** 'Potential for Plastics to Transport Hydrophobic Contaminants', E.L. Teuten et. al., *Environmental Science & Technology*, November 2007.

270 **A science writer:** 'A Toxic Trojan Horse: Tiny Plastic Particles Pack a Major pPunch', Zoe Cormier, *Globe and Mail*, 17th November, 2007.

270 **In 2012, Dutch scientists:** 'Invisible Plastic Soup is Harming Ocean Animals', Environment News Service, 26th September, 2012.

270 **A 2016 study showed:** 'Attenborough Wants Action on Plastic in Sea', Ben Webster, *The Times*, 4th February, 2016.

270 **The fish also ignored:** 'Microplastics Killing Fish Before They Reach Reproductive Age, Study Finds', Fiona Harvey, *Guardian*, 2nd June, 2016.

271 **It also became apparent:** 'Assessing the Release of Microplastics and Nanoplastics from Plastic Containers and Reusable Food Pouches: Implications for Human Health', Kazi Albab Hussain et al., *Environmental Science & Technology*, 21st June, 2023.

271 **The microbeads became a priority:** Author interview with Gavin Warner.

271 **Studies showed that plastic recycling plants:** 'The Potential for a Plastic Recycling Facility to Release Microplastic Pollution and Possible Filtration Remediation Effectiveness', Erina Brown et al., *Journal of Hazardous Materials Advances*, Vol. 10, May 2023.

271 **reusable plastic food containers:** 'Assessing the Release of Microplastics and Nanoplastics from Plastic Containers and Reusable Food Pouches: Implications for Human Health', op. cit., and 'Microplastics in Polystyrene-made Food Containers from China: Abundance, Shape, Size, and Human Intake', Jianqiang Zhu, *Environmental Science and Pollution Research*, 6th January, 2023.

271 **Researchers showed that hazardous chemicals:** 'Novel Insights into the Dermal Bioaccessibility and Human Exposure to Brominated Flame Retardant Additives in Microplastics', Ovokeroye A. Abafe et al., *Environmental Science & Technology*, 14th July, 2023.

271 **In 2022, a prominent London law firm:** 'Microplastics are a Ticking Timebomb for Litigation', RPC, published on 25th March, 2022.

272 **Then, in 2024:** 'Microplastics and Nanoplastics in Atheromas and Cardiovascular Events', Raffaele Marfella et al., *New England Journal of Medicine*, 6th March, 2024.

272 **Ours is the first evidence:** Author interview with Antonio Ceriello.

273 **plastic was the most problematic:** As described in author interview with Gavin Warner.

274 **We are inviting the entire:** 'It is Time to Recycle More Before We Drown in Plastic', Paul Polman, *The Times*, 18th January, 2017.

275 **In April 2017, Nils Simon:** Simon wasn't the first person to suggest a plastics treaty. In 2013, a paper from UCLA's

Institute for Environment and Sustainability recommended 'a new international plastic marine litter treaty of the scale and scope of the Montreal Protocol'. Simon's vision, however, was for a much broader treaty and his report got traction.

275 **Why is there no regime on plastics?:** Author interview with Nils Simon.

276 **In China:** 'China Quick to Soak Up Blue Planet', Jonathan Leake and Yuqiong Zheng, *Sunday Times*, 12th November, 2017.

276 **What I said to the crew:** Author interview with James Honeyborne.

277 **Think about parents coming back:** Author interview with Lucy Quinn.

278 **This had to happen in practice:** The EMF's suggested test to determine recyclability is whether packaging disposed of by consumers goes on to hit a 30 per cent recycling rate in all regions where it is sold or, for larger companies, across regions that collectively house at least 400 million people.

279 **it was taking 58 per cent:** Data provided by the Recycled Materials Association.

279 **But much of what the Western world:** 'Hong Kong Customs Seizes Alleged Toxic Shipment', Tan Ee Lyn, Reuters, 22nd September, 1997.

279 **Single-stream recycling:** 'After Study, Recycling Authority Decides Against Switch to Single Stream', Sarah Rafacz, *Centre Daily Times*, 7th December, 2016.

279 **On a 1992 visit:** 'Plastics: Trashing the Third World', Annie Leonard, *Multinational Monitor*, June 1992.

279 **Criticism began to mount:** 'China's Environmental Woes, in Films That Go Viral, Then Vanish', Kiki Zhao, *New York Times*, 28th April, 2017.

280 **By 2019, the US's plastic waste exports:** Data provided by the Recycled Materials Association.

280 **While previously it had sold:** 'Reduce. Reuse. Confuse.', Consumer Brands Association, 2019.

280 **We had no alternative:** 'Recycling Rethink: What to Do With Trash Now That China Won't Take It', Saabira Chaudhuri, *Wall Street Journal*, 19th December, 2019.

281 **The brands are going to have to show:** 'APR: China Ban May Boost Pressure for Recycled-Content Packaging', Steve Toloken, *Plastics News*, 28th August, 2017.

281 **The United Nations Environment Program:** 'Big Brands Pledge to Turn Tide on Global Plastic Waste', Reuters, 28th October, 2018.

285 **Only 68 per cent of countries:** Ibid.

285 **There were seven different definitions:** Ibid.

285 **A 2020 analysis showed:** 'Breaking the Plastic Wave: A Comprehensive Assessment of Pathways Towards Stopping Ocean Plastic Pollution', PEW Charitable Trusts and SystemIQ, 2020.

286 **For companies and investors:** Statement for UNEA 5.2 by business manifesto signatories, January 2022.

286 **Without global rules:** 'Businesses See Plastics Treaty as Good for Investment', Steve Toloken, *Plastics News*, 7th February, 2022.

287 **It wouldn't even:** 'ExxonMobil Investors Reject Plastic Risk Study', Steve Toloken, *Plastics News*, 3rd June, 2024.

287 **Fragmented bans weren't working:** 'Global Plastic Waste Set to Almost Triple by 2060, Says OECD', OECD, 3rd June, 2022.

287 **They can make up:** 'State of the Science on Plastic Chemicals – Identifying and Addressing Chemicals and Polymers of Concern', Martin Wagner et al., *PlastChem*, 2024.

288 **Studies have found:** 'Non-Intentionally Added Substances (NIAS) in Recycled Plastics', Oksana Horodytska et. al., *Chemosphere*, July 2020.

290 **As early as the 1970s:** 'Health Hazards: Vinyl-Chloride Risks Were Known by Many Before First Deaths', Barry Kramer, *Wall Street Journal*, 2nd October, 1974.

290 **abnormally high miscarriage rates:** 'Vinyl Chloride Tied to Higher Incidence Of Miscarriages', Barry Kramer, *Wall Street Journal*, 5th February, 1976.

290 **Studies have shown that certain phthalates:** 'Perinatal Exposure to Phthalates', Laura Lucaccioni et al., *International Journal of Molecular Sciences*, 14th April, 2021.

290 **They have been linked:** 'Dangerous PFAS Chemicals Are in Your Food Packaging', Kevin Loria, *Consumer Reports*, 24th March, 2022.

290 **In Europe in 2023:** This refers to the amount of a substance that can be ingested daily over one's life without an appreciable health risk.

291 **It sounds draconian:** 'Bisphenol S in Food Causes Hormonal and Obesogenic Effects Comparable to or Worse than Bisphenol A: A Literature Review', Michael Thoene et al., *Nutrients*, February 2020.

293 **Lawmakers continue to push:** 'California Puts EPS Foodservice Makers on Notice to Meet Recycling Targets', Steve Toloken, *Plastics News*, 12th September, 2024.

293 **claim the containers:** 'Lawmakers Seek a National Foam Ban in 2026', Maria Rachal, *Packaging Dive*, 12th December, 2023.

293 **Between 2011 and 2018:** 'Open Defecation by Proxy: Tackling the Increase of Disposable Diapers in Waste Piles in Informal Settlements', Hannah L. White et al., *International Journal of Hygiene and Environmental Health*, May 2023.

293 **The disposable diaper industry:** 'A Circular Economy for Nappies', Members of the Ellen MacArthur Foundation network, October 2020.

294 **Meanwhile, the North American market:** Data from Mordor Intelligence.

294 **Of all the plastics:** 'New Report Reveals that U.S. Plastics Recycling Rate Has Fallen to 5%–6%', *Beyond Plastics*, 4th May, 2022.

294 **A 2019 survey:** 'Reduce. Reuse. Confuse.', Consumer Brands Association, 2019. The CBA report said 68 per cent of those polled said they assumed that any product with symbols for all seven codes would be recyclable. Another 24 per cent said they didn't know. Just 8 per cent said no.

295 **Fewer than 12 per cent:** Data provided by NAPCOR for 2022.

295 **It didn't set a new one:** The Coca-Cola Company Evolves Voluntary Environmental Goals, press release, 2nd December, 2024.

Chapter 13: The Future of Plastics is Not in the Trash Can

297 **It takes 93 plastic cucumber wraps:** 'To Wrap or to Not Wrap Cucumbers?', Chandrima Shrivastava et al., *Frontiers in Sustainable Food Systems*, Vol. 6, 2022.

299 **60 per cent of food waste:** 'Food Surplus and Waste in the UK Key Facts', WRAP, November 2023.

299 **Fresh vegetables and salads:** 'Household Food and Drink Waste in the United Kingdom 2021–22', WRAP, 23rd November, 2023.

301 **more than 50 per cent of the honey sold:** 'Out of the Cracker Barrel', T. Swann Harding, *Southern Economic Journal*, October 1938.

301 **Packaging has been carried out:** Ibid.

303 **In Germany, where the polluter pays principle:** 'Current Recycling Review: Successes, Failures and Challenges', *Umwelt Bundesamt*, 4th December, 2023.

303 **Germany has recycled:** 'Indicator: Recycling Municipal Waste', *Umwelt Bundesamt*, 4th October, 2024.

304 **An example is the UK's 10-pence fee:** 'Plastic Bag Use Falls by More Than 98% After Charge Introduction', DEFRA, 31st July, 2023.

308 **Ocado has worked:** Interview with Rob Spencer of GoUnpackaged, who worked with Ocado on the refill effort.

313 **At the 2024 Olympics in Paris:** 'Paris Olympics: Plastic as Present as Ever with Coca-Cola at Olympic Venues, Despite Promises of "Green" Games', Stéphane Mandard, *Le Monde*, 4th August, 2024.

314 **It also helps that Danes:** Data is for 2023 and was provided by Dansk Retursystem. The overall return rate for PET, aluminium and glass drinks containers was 92 per cent in 2023.

318 **McDonald's sent me a study:** 'No Silver Bullet: Why the Right Mix of Solutions Will Achieve Circularity in Europe's Informal Eating Out (IEO) Sector', Kearney, 2023, https://nosilverbullet.eu/wp-content/uploads/2023/02/No-silver-bullet%E2%80%93why-a-mix-of-solutions-is-required-to-achieve-circularity-in-Europe.pdf

320 **Glass is made in furnaces:** European Climate, Infrastructure and Environment Executive Agency, 16th March, 2022.

FAQs

326 **On the other hand, single-use plastics:** 'OECD Working Paper: Preventing Single-Use Plastic Waste', 20th October, 2021.

327 **That's 70 days longer:** 'Reducing Household Food Waste and Plastic Packaging', WRAP, February 2022.

332 **The results are specific to the UK:** 'Life Cycle Assessment (LCA) of Disposable and Reusable Nappies in the UK', Giraffe Innovation, 2023.

335 **It's highly water-intensive:** 'Disposable Paper-Based Food Packaging: The False Solution to the Packaging Waste Crisis', Manon Stravens, *Profundo*, 28th July, 2023.

335 **a 2021 report found:** 'Overview of Intentionally Used Food Contact Chemicals and Their Hazards', Ksenia J. Groh et al., *Environment International*, Vol. 150, May 2021.

337 **Solvent-based processes:** 'Plastic Recycling Reimagined', Adam Herriott, WRAP, October 2023.

338 **By one estimate:** 'Breaking the Plastic Wave: A Comprehensive Assessment of Pathways Towards Stopping Ocean Plastic Pollution', PEW Charitable Trusts and SystemIQ, 2020.

339 **the majority of kerbside-collected food waste:** REA written evidence to EFRA committee inquiry on Plastic Waste, 21st February, 2022.

339 **In the US:** Data from Bryan Staley of the Environmental Research and Education Foundation.

339 **Not one indicated:** 'Composting State of Practice: Results from a National Operations Survey', Environmental Research & Education Foundation, 2024.

340 **While some proponents:** 'Relevance of Biodegradable and Compostable Consumer Plastic Products and Packaging in a Circular Economy', *Eunomia*, March 2020.

340 **It can take months:** Ibid.

340 **Brands also promise:** Interview with Ramani Narayan, a professor at Michigan State University who has been researching biodegradable plastics for more than 30 years.

342 **About 48 per cent of bioplastics:** European Bioplastics.

342 **Studies show that some:** 'Energy Analysis of Bioplastics Processing', Christine Schulz et al., *Procedia CIRP*, Vol. 61, 2017.

342 **Pesticides, PFAs and other chemicals:** 'Towards Safe and Sustainable Food Packaging', BEUC, 2021.

342 **The electricity generated:** 'Debunking Efficient Recovery: The Performance of EU Incineration Facilities', Zero Waste Europe, January 2023.

343 **These are linked:** 'Breaking the Plastic Wave: A Comprehensive Assessment of Pathways Towards Stopping Ocean Plastic Pollution', op. cit.

343 **That view is starting to change:** 'Microplastics in Landfill Leachate: Sources, Detection, Occurrence, and Removal', Mosarrat Samiha Kabir et al., *Environmental Science and Ecotechnology*, 16th February, 2023.

344 **concluded that the additives:** 'Transport and Release of Chemicals from Plastics to the Environment and to Wildlife', Emma L. Teuten et al., *Philosophical Transactions of the Royal Society B: Biological Sciences*, 27th July, 2009.

345 **I have replaced:** 'From E-Waste to Living Space: Flame Retardants Contaminating Household Items Add to Concern About Plastic Recycling', Megan Liu et al., *Chemosphere*, October 2024.

346 **Researchers have also raised concerns:** 'Health Effects of Microplastic Exposures: Current Issues and Perspectives in South Korea', Yongjin Lee et al., *Yonsei Medical Journal*, May 2023.

347 **For instance, a 2024 study:** 'Influence of Colourants on Environmental Degradation of Plastic Litter', Sarah Key et al., *Environmental Pollution*, 15th April, 2024.

ABOUT THE AUTHOR

Saabira Chaudhuri has covered consumer goods companies for the *Wall Street Journal* for the past decade and has reported on plastics from the US, India, UK and elsewhere in Europe. She has an MA in business and economic reporting from New York University and a BA in sociology from Mount Holyoke College. Saabira lives in London with her husband, two children and dog. She grew up in Bangalore, India, where she first developed her fascination with what we throw away.